THE
BASINGSTOKE & ALTON
LIGHT RAILWAY

An Edwardian view of Bentworth, centred on the village pond looking towards the Star Inn and the village green. Communities such as this could not be expected to generate much traffic on the newly opened railway.

THE

BASINGSTOKE & ALTON

LIGHT RAILWAY

Martin Dean
Kevin Robertson
Roger Simmonds

Crusader Press

Dedicated to
the memory of the late Edward Griffith

Edward Griffith on the footplate of 'Q' class No. 545 at West Meon in July 1947, two months before the publication of the first booklet on the branch line.

First published 1998 by Crusader Press

ISBN 0 9534197 0 3

Typeset by Highlight Type Bureau Ltd, Bradford
Printed in Great Britain by The Amadeus Press Ltd
Huddersfield, West Yorkshire

CRUSADER PRESS
'Narnia', Flaxpool, Crowcombe, Somerset TA4 4AW

CONTENTS

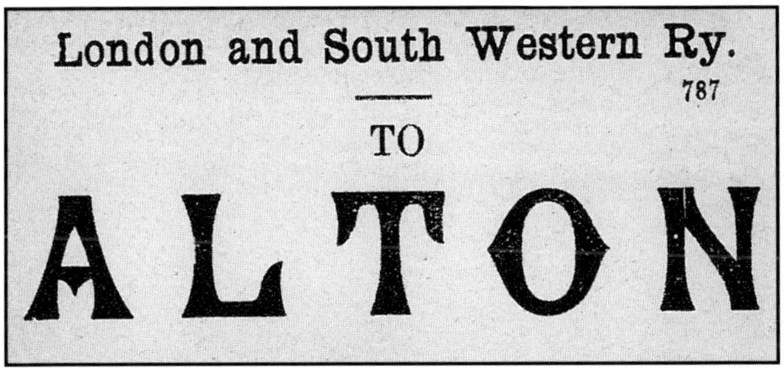

London and South Western Ry.
787
TO
ALTON

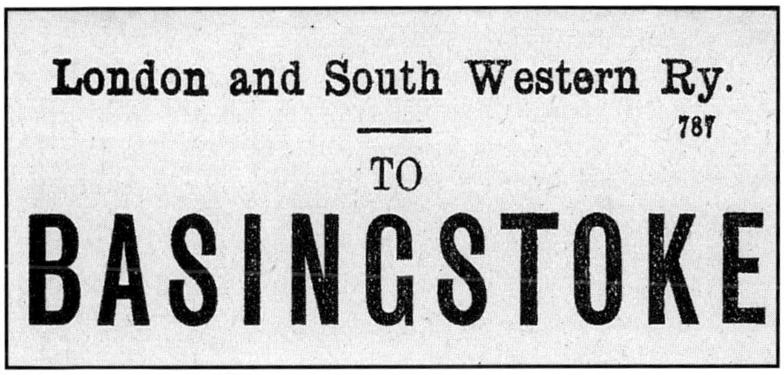

London and South Western Ry.
787
TO
BASINGSTOKE

PREFACE

How many parents face the problem of "entertaining" their children through the seemingly unending summer holidays? The parents of one of the authors certainly did. He well remembers one occasion in 1959 when he was in his early teens. His mother in desperation, suggested he went to the Classic Cinema, South Croydon, where Will Hay's comedy "Oh, Mr. Porter!" was showing. He was unimpressed with this suggestion, the film being over twenty years old and unknown to him. Only when the financial incentive was offered was the decision to go made. How could he refuse?

So it was he made contact with the Basingstoke and Alton Light Railway, albeit unknowingly. He thoroughly enjoyed the film and it remained one of his favourites for years. Because of this he became interested in the film's background. Much later he found out where it had been "shot" and something of the peculiar history of the railway line in question. This book endeavours to trace that history and attempts to answer some of the questions that still surround it one hundred years after its construction began.

INTRODUCTION

The immediate post-war years saw a new trend - the publication of booklets on railway branch lines. Invariably published privately, their print runs were low and their availability limited at a time when rationing was still in force. One such was by Edward Griffith of Farnham, Surrey who in September 1947 published an illustrated booklet on the Basingstoke and Alton Light Railway. Very much an obscure branch the line had been closed for some years, but nevertheless Mr. Griffith set out on his task with great tenacity, in the course of which he was able to interview former members of staff. Reprinted and largely rewritten in 1970 the booklet has been out of print for many years. With new research material being made available the time has come to produce a new volume on this railway byway. Its history is certainly peculiar and this makes for a fascinating story, full of conflict.

Every railway line is unique, but as is well known some are more noteworthy than others. Without a doubt the Basingstoke and Alton Light Railway is worthy of special mention. It was the first line authorised under the 1896 Light Railways Act and the reasons for its construction have always been in doubt. For many years it was thought that it was built purely as a blocking line; a small part in the strategy of railway politics played so wastefully by the railway companies in the latter 19th century. This volume casts doubt on this traditional view.

However, if its pre-history has been subject to conjecture there can be no doubts surrounding its subsequent fate. Here it is totally unique for no railway before (and we believe since) has been abandoned, dismantled and rebuilt the latter as a result of public pressure and against the wishes of the railway company. Nevertheless within thirteen years of its re-opening, the line was closed for a second and final time. Yet, within this period the branch was to feature in two full length films so that even today the line can be seen in part, sixty years after the last train ran.

In attempting to chronicle the history of the branch the authors have portrayed the line within the living landscape. They have tried to ascertain its role in the local economy and the reasons for its failure.

We hope you enjoy reading this book - certainly the authors have derived great pleasure in both researching and writing it.

CHAPTER 1
EARLY SCHEMES AND EARLY DREAMS

Geographically, the area between Basingstoke and Alton is unpromising railway territory. It consists of a plateau of high chalk downland with much land in excess of 500 feet above sea level. The pervious nature of the soil meant that natural water supplies have always been a problem, it often being necessary to sink wells to a depth of over 300 feet. This, coupled with the relatively poor soils, meant that the population density was low, being mainly located in small scattered villages.

In the 19th century, agriculture was the predominant industry and was organised in large estates with numerous tenant farmers. The chalk soils supported traditional crops such as wheat and barley whilst in the area around Alton, hops were grown in quantity.

The upland country took the form of a series of rounded rolling hills these often being covered with woods of oak and ash. The undergrowth tended to be of hazel and this was used in the making of wattle hurdles. Where the clay covering was thinner beech woods were more common, but overall the landscape was open and bare. It has been described as an area of "cheap and dry land". Uncompromising territory indeed.

Communications tended to be poor, although the locally important market town of Basingstoke was linked to Reading by turnpike road as early as 1718. Further routes followed but it was not until 1795 that Basingstoke was linked to Alton by turnpike. Even so the route was beset by steep hills as it climbed out of the valleys in which Basingstoke and Alton lay. With poor traffic potential it is not surprising that Hampshire's turnpikes had one of the lowest earnings per mile in England.

Basingstoke's economy flourished in the 18th century as it became increasingly important as a market centre. The corn trade was especially valuable as was the local manufacture of woollen textiles. To facilitate the town's prosperous trade the Basingstoke Canal was built. Opened on 4th September 1794 it encouraged the wider and cheaper distribution of the area's agricultural produce whilst also facilitating the cheaper supply of coal, iron and bricks. The town was now poised for expansion and this can be seen in the census returns. In 1801 the population stood at 2,589 - one hundred years later the figure was 9,510. The manufacturing base had totally changed with the production of Wallis and Stevens traction engines and soon, Thornycroft lorries.

It was into this agricultural economy that the railway companies laid their first tentative lines. Basingstoke was placed on the railway map on 10th June 1839 when the London and Southampton Railway opened its route from London. The following year the undertaking was completed and opened throughout. The Basingstoke Canal which largely paralleled the railway suffered a sharp reduction in trade immediately the railway opened and went into terminal decline. Traffic into Basingstoke seems to have ceased by 1910.

The pioneering railway years were marked by rival companies promoting competing routes. In an era of free enterprise and lack of Government interference this was perceived as acting in the public interest. So it was the London and Southampton Railway, soon to be re-styled the London and South Western Railway (LSWR), cast their eyes towards Bristol and proposed a "branch" from Basingstoke to that city. The proposal failed and Brunel's rival Great Western Railway (GWR) was built between the capital and the West's premier city. Consequently open rivalry existed from the start between the two railway companies as they sought to out-manoeuvre each other.

In 1848 the GWR penetrated its rival's 'territory' with the opening of the Berks and Hants branch from Reading to Basingstoke. Only eight years later in 1856 they reached Salisbury from Westbury. Nevertheless it was the Berks and Hants route that was to be the most important as it provided the impetus for later schemes to link the Midlands and the North to Portsmouth via the GWR.

Whilst these extensions were coming into use the LSWR had not been idle. Salisbury was reached via Bishopstoke (Eastleigh) and Romsey in 1847 whilst just west of Basingstoke a junction with the main line led to a separate route to Andover. Alton was reached from London via Farnham on 28th July 1852 and this line was extended through the Hampshire downlands and along the upper reaches of the Itchen valley to Winchester. Known as the Mid Hants Railway the part between Alresford and Alton survives today as the 'Watercress Line'.

Thus by the latter part of the 19th Century the LSWR had established a virtual monopoly of railway carriage in Hampshire with a network of routes serving the main centres of commerce, industry and population. However competition existed in certain locations; at Winchester and Basingstoke with the GWR and at Portsmouth with the London Brighton and South Coast Railway (LBSCR). Nevertheless the most important town, Southampton, remained a LSWR monopoly. Even so there were continuous attempts to wrest this lucrative traffic away, especially by the GWR. The closest they got was in 1882 with their support for the nominally independent Didcot, Newbury and Southampton Railway. Opened in 1885 it was worked by the GWR from the outset. However, the line only reached Winchester, twelve miles away from Southampton. Financial insolvency ensured that the line did not reach its projected destination and it had to be content with a junction with the LSWR main

Basingstoke station looking west before rebuilding. To the left are the original LSWR 'up' and 'down' platforms and in the centre the 'up' bay. The GWR station is enclosed within the overall roof to the right. Concurrent with the building of the light railway was a major redevelopment of the LSWR station which was itself associated with the quadrupling of the main line. Accordingly the former down platform on the left became an island and the buildings were demolished. Two further sets of rails and the associated platform were then added to the extreme left. *(Lens of Sutton).*

line near Shawford, just south of Winchester.

An earlier attempt to reach Southampton had been made by the LBSCR in 1866. They put forward proposals for a Midhurst-Petersfield-Bishops Waltham-Southampton route but the plans were thwarted by the LSWR opening its own Petersfield to Midhurst branch in 1864. The financial and banking crisis which hit the country in 1866 created chaos and confusion and spelt the end of numerous speculative railway ventures. The LBSCR was badly affected (it stopped building its Ouse Valley line) and made no further attempts to expand into Hampshire.

Yet another potential competitor was the South Eastern Railway (SER). Opening the first section of their Reading to Redhill route in 1849 the company had unsuccessfully attempted a take over of the Portsmouth Direct Railway from Guildford. Subsequently their main area of concern was within Kent and the company did not pose a direct threat to the LSWR. However always seeking to expand the traffic on their cross-country route, they were to show they would support any opportunities that were presented.

What all this shows is that despite having established a dominant position within Hampshire, the LSWR did not have the situation totally under control. It was a matter of constantly being aware of rival proposals and then taking the necessary steps to maintain its commercial advantage. Nevertheless the LSWR always saw the GWR as its biggest threat and hence schemes obtaining the support of that company (directly or indirectly) were always viewed with the deepest suspicion by the LSWR Board.

By the late 19th Century an element of business maturity had developed between the railway companies. Returns on capital were reducing whilst operating costs were increasing. Structural rigidity necessitated commercial re-appraisals. As early as 1862 the LSWR and the LBSCR entered into an agreement whereby the Portsmouth traffic was pooled and the receipts shared on a fixed percentage basis. This comfortable duopoly no doubt benefited the railway companies but it led to dissatisfaction within the town. Consequently from the 1880's various attempts were made to link Portsmouth to the GWR. This was poor timing by the Portsmouth commercial interests as the

Some of the Basingstoke staff pose outside the down side station buildings in 1898. At least 38 railwaymen are visible although this is less than the total employed at the location as clearly no locomotive department staff are present.

GWR and LSWR chose to enter into a private agreement over territorial boundaries in 1884. The reason for this has been explained by the GWR's wish to maintain existing boundaries and the fear that the LSWR might be encouraged to extend to Bristol as a retaliatory measure. Naturally this agreement was not made public but it would explain why the variety of schemes promoted to serve the area between Basingstoke and Alton were all doomed to failure. Essentially they were all planned as through routes with little consideration for the local traffic; this was always peripheral to the overall strategy which incorporated somewhat grandiose schemes to link Portsmouth to the Midlands.

The first suggestion for a railway joining Basingstoke and Alton can be found in an 1884 proposal for a Basingstoke, Alton and Petersfield Railway. It is believed that a group of London financiers were behind the scheme acting through their solicitors Messrs Soames, Edwards and Jones. The railway would have left the LSWR at Basingstoke and for several miles have followed a south-easterly course similar to that built later. Approaching Alton it would have crossed the Mid Hants line at right angles before continuing in a south-easterly direction along a tortuous route via Farringdon, Newton Vallance and Steep, terminating in an end on connection with the LSWR Portsmouth direct line just north of Petersfield. A further connection would have provided a direct link to the LSWR Midhurst branch. Traffic from Alton would have been able to join the new line via a connection facing Petersfield, the intention being to tap the valuable military traffic between Aldershot and Portsmouth. Running powers over the GWR from Reading to Basingstoke, the LSWR from Petersfield to Portsmouth and into the LBSCR station at Midhurst were also sought, although a somewhat surprising omission was that over the LSWR Midhurst branch.

The Engineer for the proposal was George Wells from the Westminster firm of Wells, Owen and Elwers. This survey of the route proposed a ruling gradient of 1 in 100 although around the appropriately named location of Steep near Petersfield, this would have been very difficult to maintain without recourse to extensive earthworks. The cost of constructing the 22 mile 3 furlong line was estimated at £450,000.

The proposal was submitted to Parliament in early 1884, with Mr Bidder QC being the principal witness in

Petersfield station in February 1911 looking north to Guildford and London. The line to Midhurst branched to the right beyond the level crossing. It was to here that the earliest proposals for a railway from Basingstoke were based.

Brewing was an important industry in Alton based on the production of locally grown hops. This led to considerable freight traffic. In this 1899 view a road wagon of Messrs. Trimmer piled high with hop pockets is standing outside the station goods yard. In the background it is just possible to discern the office of Sayers' - coal merchants. The fine condition and decoration on the horses shows how important they were to business in the pre-motor era.

favour of the Bill. Interestingly the GWR declined to appear and were recorded as being against the proposal. The conflicting arguments were eventually referred to committee where the proposal failed, the Basingstoke, Alton and Petersfield being lost forever.

The failure of this first scheme was not really surprising. The idea of tapping the military traffic would not meet with the support of the LSWR who probably felt that it was already adequately served by its existing lines. It was also totally irrelevant to the section between Basingstoke and Alton. However the opposition of the GWR, who probably feared the running powers over its Reading line, was the final straw. With no support from the two main line companies failure was inevitable.

Undeterred Soames, Edwards and Jones tried again four years later. In their 1888 proposal for a Basingstoke, East Hampshire and Portsmouth Railway, they planned a direct line to Portsmouth ending at a terminal adjacent to Southsea Pier. A branch was to serve Havant joining the LBSCR by a triangular junction. Running from the GWR station at Basingstoke the plans envisaged a junction with the LSWR Mid Hants line near Alton before continuing down the Meon Valley.

Soames, Edwards and Jones sought the services of Alexander Downie a well established solicitor in Alton. No doubt they felt his presence would encourage investment from the area the line was to serve. This appears to be the first instance that Downie was connected with railways in the Alton area. In later years in his capacity as Clerk to Alton Rural District Council (RDC) he was to have many dealings with the Basingstoke and Alton Light Railway. On this occasion his attempts were not successful and the proposal, like its predecessor, failed to attract sufficient support.

There seems to have been little in the way of action during the next few years although it is obvious that there were interests in Portsmouth still keen to link that city with the GWR. A recently discovered file of correspondence shows that the 'Portsmouth Railway Extension Association' made an approach to the Didcot Newbury and Southampton railway for a line from Winchester to Portsmouth via Botley Common. The line which would have been 24 miles long was mooted in late 1894 by a DNS shareholder who was also a member of the Association. The DNS, with problems in completing its own route, showed no interest. Nevertheless the Association were still pursuing their objective by whatever means possible and suggested a line linking Portsmouth with Basingstoke "so as to afford direct access to the GWR." Clearly some of the influential citizens of Portsmouth were totally dissatisfied with their railway companies; the LSWR and LBSCR having become complacent, with their pooling arrangement obviously working to their advantage - not their customers'.

Staff at Alton station probably taken in the 1890's. Sitting to the right of the Station Master attired in a bowler hat is the chief clerk.

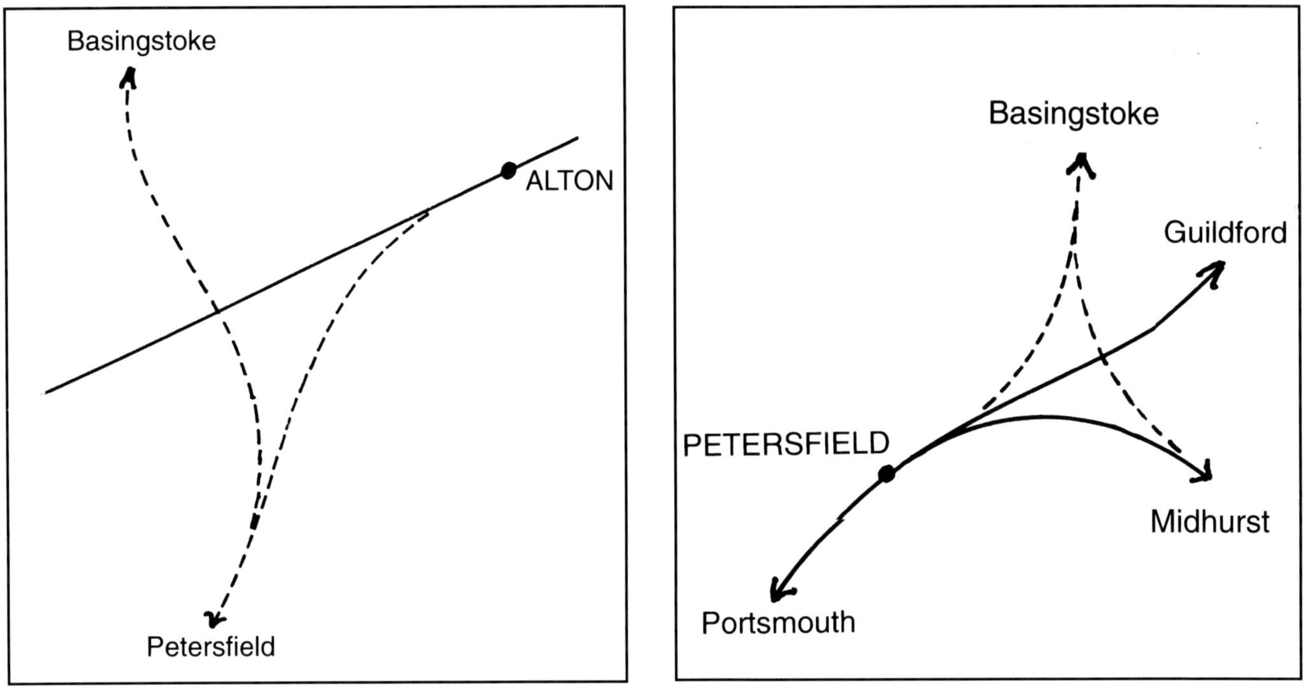

Sketches to show the 1884 proposed railway scheme with its links at Alton and Petersfield.

Compared to thirty years earlier, the mid 1890's were a period of considerable prosperity for the railway industry and this, coupled with the continuing dissatisfaction with the railway service at Portsmouth, led a group of businessmen from the Portsmouth area to actively pursue the idea of linking their city to the GWR. Led by the Mayor and John Brickwood, brewer and President of Portsmouth Chamber of Commerce they formed a Portsmouth Meon Valley and Alton Railway Company. The proposal was for a main line between Basingstoke and Portsmouth with a branch from the SER at Shalford near Guildford. This would have joined the main line at East Tisted in the Meon Valley. The promoters thus saw their escape from the LSWR/LBSCR duopoly and despite traversing through established LSWR territory no connections with that company's metals were planned.

Proposed in 1895 only one year after the idea of a DNS connection, the new line was given the heading of the Portsmouth, Basingstoke and Godalming Railway. The promoters did not foresee any undue difficulty in attracting capital from their own area although it was recognised that support from the northern part of the route was weak. Wisely they sought the services of the indefatigable Downie because it was hoped he could enlist the support of influential local landowners. His efforts were not successful for on 26th November 1895 he wrote to Basingstoke RDC stating that the Directors of the undertaking had "received no support from any part of the county north of Alton". He also went on to request that the Council "assist in the expenses of the promotion of the railway". However support was forthcoming south of Alton for one of the Directors was W.G. Nicholson, the prominent landowner of Basing Park, Privett in the Meon valley who became MP for Petersfield in 1897. Nicholson was to feature in the activities surrounding the Light Railway for many years.

Certainly the projected railway was finding the securing of capital difficult and it is likely that Downie's pleading letter to Basingstoke RDC was only one of many. The extent of the difficulty can be judged by the fact that when the proposal was presented before the House of Lords Select Committee the Company had failed to raise its £20,000 nominal capital. This did not bode well for a scheme costing £2.4 million.

The Portsmouth, Basingstoke and Godalming Railway (PBGR) Bill eventually came before Parliament in 1896. Not surprisingly it was vigorously opposed by both the LSWR and the LBSCR, the former discussing the project at the Board Meeting on 12th December 1895. Indeed a total of thirteen petitions were lodged against it, so it is understandable that the Bill failed. The LSWR and LBSCR based their opposition on the fact that Portsmouth was already adequately served by rail and that the new line was an unnecessary cost. With the opposition led by Sir Charles Scotter, General Manager of the LSWR, the proposal was rejected. Before dismissing the case it is of vital importance to highlight one point referred to by Scotter before the House of Lords Select Committee. When addressing them he called the proposed railway "a direct attack on the LSWR and the Brighton." When asked what would be a proper alternative for the Meon valley area he replied "a light railway; no other railway could pay." He went on to say that it was "a district eminently suited for an experiment to be made." When asked whether the LSWR would be prepared to act on it he replied that he "told the President of the Board of Trade that months ago." Clearly it will be seen that Scotter's comments and promise concerned the area south of Alton; nothing was said about the area to the north. Bearing this in mind it makes the subsequent developments even more surprising, for what he said to the Lords and what the LSWR did in practice were quite different.

At the Board Meeting of 28th May 1896, it was formally reported that the PBGR Bill had been rejected by the House of Lords Committee. One month later on 25th June 1896 Scotter presented a report to the LSWR Board in which he proposed a conventional railway down the Meon valley and a Light Railway from Basingstoke to Alton after the passing of the Light Railways Bill now in Parliament. The report led to much Board Room debate. As a result the Chairman W.S. Portal and Scotter inspected the district and met the important landowners, including Nicholson. On the basis of this visit the decision to build both lines was made. Unfortunately the report of the visit has been lost for it would make interesting reading and may even identify the reasoning behind the decision.

It has been said that the LSWR planned the Light Railway as a "blocking line"; an attempt to keep the GWR out of its territory. There is no evidence that this was so. All through the proceedings of the 1895 Portsmouth, Basingstoke and Godalming Railway Bill the GWR had remained silent. The failure of the proposed railway to find its nominal capital shows that there was no financial involvement, either directly or indirectly. All three railway proposals had emanated from Portsmouth with a firm desire to improve its links to the north; interest from the GWR was always negative. So why did Scotter change his mind? Probably the LSWR recognised that had the GWR supported the Bill the outcome could well have been very different. It recognised that three unsuccessful attempts had been made to penetrate its territory by main lines and that the only permanent solution to a recurring problem was to promote its own lines. The landowners obviously painted a rosy picture as to the traffic potential of the area - history was to prove them woefully wrong. The Meon valley line clearly gave the LSWR a more direct access to Gosport and could be seen as an alternative wholly owned route to the Portsmouth area, necessary if the pooling arrangements became soured. Such a positive statement could not be said for the light railway. Consequently the two new lines were soon to prove liabilities rather than assets, dubious investments indeed.

An early stage in the redevelopment work at Basingstoke station. Compared with the previous photograph it will be seen that the chimney stacks have already been demolished. The photograph belongs to another age when Basingstoke was a country market town not the modern centre it is today. The horse drawn cart and the sign on the 'Junction Inn' point to the importance of traditional horse power to deliver goods locally. Just visible under the sign a uniformed member of staff seems to be heading for the public bar.

THE RAILWAYS OF CENTRAL SOUTHERN ENGLAND.

CHAPTER 2
THE ADVENT OF THE LIGHT RAILWAY

On 14th August 1896 Royal Assent was given to the Light Railways Act (59 and 60 Vic. cap. 48). In content the Act removed certain of the restrictions previously placed upon the builders and operators of minor railways thereby making it easier for the promotion of such schemes to the obvious benefit of the communities concerned. A major concession was that approval of these lines would be given by a body called the Light Railway Commissioners who would alleviate the otherwise time consuming and expensive business of an Act of Parliament, previously the only authority for building a new railway line. In return the Light Railway Commissioners were given powers to place restrictions upon the promoters of light railways with regard to weight limits and speeds.

The intention of the Government was not specifically in the promotion of secondary lines of railway, it was intended as a relief measure to assist agriculture which was suffering from economic decline. The Queen's speech to Parliament in February 1896 made the position clear when she said that "the condition of agriculture is disastrous beyond any recent experience. Measures will be laid before you, of which the object will be to mitigate the distress under which the classes labour who are engaged in that industry." A study of farm rents in the Herriard and Lasham area shows what local conditions were like. In 1888 rental income was £2,070, by 1896 it had dropped to £1,785. At a time when direct political interference in market conditions was seen as being unacceptable, or indeed undesirable the Government saw its role as being a facilitator. Nevertheless it was also aware that lower food prices aided the urban majority. Its view was that the rural problem was partially due to inadequate transport preventing agricultural produce from reaching a saleable market and so it committed itself to what can only be described as a peripheral activity. It must be remembered that this was at a time when the railway industry was the prime carrier with little competition, so the Government's response can be seen to be logical.

The Government had cast their eyes towards Belgium and Holland where the equivalent of light railways already existed. These produced an average return of 3% per annum - a most satisfactory figure for the period when there was nil inflation and many raw material prices were actually falling. Nearer to home in Ireland the 1883 Tramway and Public Companies Act had previously permitted the building of light railways of the type now proposed. The Irish venture was vastly different to the experiences in Europe with the new railways showing little if any return and with no apparent improvement in farm prices to the communities they served.

That there was a problem in rural areas was not in doubt. Wheat prices which had averaged 37/- a quarter in 1891 fell to 26/4d in 1893 due to over production on the world market. The depression was described as "intense" with large tracts of land left uncultivated. In Hampshire many tenant farmers had given up and there was a continuous decline in the intensiveness of farming and capital investment. In northern Hampshire the Royal Commission on Agriculture found rents had been reduced by a minimum of 25%.

The Light Railways Bill was introduced into Parliament by the President of the Board of Trade, Charles T. Ritchie on 20th February 1896. He admitted that the Bill was "not a panacea for agricultural depression" but that its aim was "to bring consumer and producer closer together". The MP for Basingstoke, A.F. Jeffreys was clearly in favour of the measure and he called the Bill "an attempt in the right direction." However, he did strike a prophetic note of caution when he said that light railways "would never pay unless they were made feeders to the great (railway) systems." Opposition to the Bill was limited and concentrated on the view that railways would be built where they were not wanted and that they would be a drain on the Exchequer.

A leading article in Herepath's Railway Journal called the Bill a "moderate, practical measure" which "deserves encouragement and support". Consequently the Bill passed on to the Statute Book with limited delay. Ironically the legislation that followed it was that facilitating the use of motor vehicles on highways. The effect of the latter could not have been foreseen although the view that "there is much yet to learn about motor cars and the experience will be dearly bought" was somewhat prophetic.

The Light Railway Commissioners advised applicants of the procedure necessary for the opening of a line of railway in early October 1896. The LSWR Board, who formally decided to build a light railway between Basingstoke and Alton at their meeting on 1st October 1896 were quick to react. On 20th November the application was advertised and printed whilst the formal application for permission to build the line seeking powers "authorising construction, maintenance and working of a light railway from Basingstoke to Alton" was lodged on 31st December 1896. The Board of Trade replied that they were looking favourably upon the application and on 19th January 1897 approval was given for plans and specifications for the new line to be drawn up. Later that month the Commissioners confirmed that a Public Enquiry into the scheme had been set for 11.30 a.m. on 4th February in Basingstoke Town Hall. After years of failure the speed of action was now in noticeable contrast.

The detailed plans had been drawn up by Messrs

The Station Master's house, part of the original Alton station which in 1930 was reported to be 'one of the finest buildings in existence'. Unfortunately the building has been demolished. The railway had reached Alton from Farnham in 1852 and the station remained a terminus until the opening of the Mid Hants line to Winchester in 1865 when the present station was built.

William R. Galbraith and E. Andrews, consulting engineers to the LSWR. Dated 29th December 1896 it is obvious the LSWR were vigorously pursuing the scheme before they had received confirmation from the Board of Trade. The plans show the course of the railway largely as that built later. A curiosity at Basingstoke shows a burrowing junction thereby allowing direct access to both the up and down sides of the main line facing London. Costs for the railway's construction were estimated at £66,714. 16s 10d.. This included an allowance of £10,000 for land purchases.

The Public Enquiry was well attended. Amongst those invited to attend were the local authorities through whose area the proposed line would pass as well as local notables. These included the M.P. A.F.. Jeffreys, who had been so supportive of the Light Railways Act, and the landed interests, notably Lord Portsmouth and F.M.E. Jervoise of the Herriard Estate.

The enquiry heard evidence that the LSWR were building the line "for agricultural purposes", a report submitted showing the population of the area between Basingstoke and Alton to total only 2,400. However the census returns of 1891 show this to be an exaggeration unless the writer of the report expected the line's catchment area to be considerable.

Sir Charles Scotter gave evidence for the LSWR where he re-iterated his promise that the company would build a light railway in the Meon valley south of Alton. He specifically made the point that he had not pledged himself to a line between Alton and Basingstoke. In his evidence he stated "the Chairman of the company, Mr. Jervoise, Mr. Jeffreys and myself went over the district and I think we all came to the conclusion that an ordinary railway could not be financially successful and that a light railway if made by independent parties would not be remunerative". Nevertheless he concluded that "if it were built and worked by the LSWR it would give an immense advantage to the district and act as a feeder to the main line and that trains would in every instance run in conjunction with those to Waterloo, Southampton, Salisbury and Andover as well as other stations on the main line." Because of this he hoped landowners would be content with parting with land at agricultural values and that the line would "lead to the rapid development of the neighbourhood." Support for the new line was almost unanimous, with only a few adverse comments. John G. Wood of Alton RDC objected to the position of level crossings on district roads whilst Lord Portsmouth suggested that the proposed Cliddesden Station should be positioned at a more convenient site to the village. These arguments would surface again many times.

The very first volume of the 'Railway Magazine' carried an interview with Sir Charles Scotter in which reference was made to the proceedings:-

A panoramic view of Alton just to the west of the station. In the foreground is the Mid Hants line which was doubled with the opening of the lines to Basingstoke and Fareham. The date is circa 1895.

The area was rural and local industry tended to reflect the agricultural base. Timber and timber products were always important and this turn of the century view shows a hurdle maker at work.

"The Commission under the statute had no sooner been created than Sir Charles, dispensing altogether with the aid of counsel or solicitors, appeared before the Earl of Jersey and his colleagues at Basingstoke to ask that the LSWR might be authorised to construct by means of a single line, about twelve miles long and of 4ft $8^1/_2$ins gauge a "light" railway running from Basingstoke through a rich farming district and forming a connection at Alton with the Company's Alton to Winchester line and at Chawton close by a junction with the new line to Portsmouth."

"Sir Charles Scotter personally conducted the case in support of the proposed line tendering his own and calling other evidence as to its urgent necessity in order to cheapen the transit of agricultural produce; and he further rebutted the few objections which were urged against the scheme. The enquiry, which naturally aroused a good deal of local enthusiasm was conducted in a little more than a couple of hours, and the Earl of Jersey as Chairman, in announcing the decision of the Commissioners in favour of the necessary consent being given for the construction of the line, expressed satisfaction that the first light railway sanctioned under the Act of Parliament was to be made under such favourable auspices as those of the LSWR Co."

After such a favourable Public Enquiry it would be natural to assume that it would only be a matter of weeks before formal authorisation was given to proceed with the works. This was not to be the case for a tentative enquiry by the LSWR on the 18th February revealed the matter was still being considered by the Commissioners. Provisional approval was given on 22nd July but it was not until 9th December 1897 that the formal Light Railway Order was issued giving powers to the LSWR to acquire lands and construct the line. Ten months had elapsed and the initial euphoria seems to have evaporated somewhat. Coincidentally, on the same day that the Board of Trade authorised the line, Sir Charles Scotter's retirement was announced. His replacement was Charles Owens, the Assistant General Manager.

Why then the delay in authorising the route, especially when the intention of the Light Railway Act had been to reduce bureaucracy to a minimum? It is believed the answer lies with the two objections originally raised at the Public Enquiry - the issue of the location of Cliddesden Station and the use of ungated level crossings with cattle grids.

The Light Railway Order has some interesting features and its contents can be found in the "Railway News" for 1898. The first was that three stations had to be provided, although timetables need not displayed. Platforms had to be built unless carriages had proper means of access to ground level. Surprisingly despite

An Edwardian Hampshire scene. The view down Drury Lane towards Ham Farm, Bentworth.

the requirement for platforms, there was no obligation to provide shelters or conveniences. Signalling details were specific. Where passing places existed, a home signal for each direction had to be provided, and if it could not be seen for 1/4 mile, had to be preceded by a distant signal. Home and distant signals could be worked from the stations by wires, but were to be weighted so they flew to danger if the connecting mechanism broke. The maximum speed limit on the line was fixed at 20 m.p.h. which was to be reduced to 10 m.p.h. on curves of less than 9 chains radius. The Company was to complete the works within three years, failure rendered them to a fine of £20 per day unless they were prevented "by unforeseen accident."

Galbraith and Andrews' engineering survey and report made reference to the burrowing junction at Basingstoke but Scotter's comments regarding a junction with "the new line to Portsmouth" (the Meon Valley line) seem to have quietly died. The costs make interesting reading and are itemised below:-

Railway No.1 Length 12m 4f.	**£**
Earthworks - rock	nil
Embankments and cuttings:	
chalk; 230,000 cu. yds. @ 1/- per yard	11,500
roads 2,000 " " @ 1/- " "	100
Bridges - public roads (8 in all)	5,200
Accommodation bridges and works	5,000
Viaducts	nil
Culverts and drains	1,250
Metalling, fencing of roads and level crossings	550
Gatekeepers houses at level crossings	1,200
Permanent Way (incl. fencing) @ £1,750 per mile	21,875
Permanent Way for sidings and junctions	2,000
Stations	3,000
Sub total	51,675
Contingencies	2,583.15/-
	54,258.15/-
Land and buildings (100 acres)	10,000.
Total	**64,258.15/-**

Railway No. 2 Length 3f 3 1/2 chains	**£**
Earthworks:- chalk and soft soil 3,000 cu. yds. @ 1/- per yd	150
Accommodation bridges	300
Culverts and drains	50
Permanent Way (incl. fencing)	732.16.3
Permanent Way for sidings and junctions	1,000
	2,232.16.3
Contingencies @ 10%	223. 5.7
Land (included in Railway no.1)	nil
Total Cost	**2,456.1.10**

TOTAL COST RLYS. 1 AND 2 £66,714.16.10
(say £67,000)

Railway No.1 was from Butts Junction, Alton to the 'up' side at Basingstoke whilst railway No.2 was from the 'down' side looping round into the single line. Interestingly the contingency figure for Railway No. 1 was only 5% whereas 10% was allocated for Railway No. 2.

Broad Walk, Hackwood - the home of the Earl of Portsmouth. This estate dominated the local economy of Cliddesden.

Capital to build the line was fixed at £51,000 with powers to raise a further £17,000 on debentures. In the event funding was undertaken under the LSWR (Additional Powers) Act of 1897 which authorised the raising of £600,000 for general line improvements as well as the £51,000 for the new line. Little trouble was experienced, the Company being viewed as a steady, sound investment and by April 1899 all £651,000 had been subscribed.

With permission to proceed, matters began to move again. Mr. M. Wallington was appointed surveyor and Galbraith the consulting engineer. Although only a single line of rails were to be laid, enough land was purchased for a double line showing a certain amount of wishful thinking if nothing else. Staking out the route commenced in March 1898 with the task completed by 6th June. Meanwhile Galbraith had been working on the engineering specification and on his recommendation tenders were sent to ten contractors. It would appear that by this time the junction arrangements at Basingstoke had been simplified and the junction to the 'upside' had been deemed unnecessary. Eight contractors replied and on 6th July 1898 the LSWR Engineering Committee considered the following quotations:-

	£
John Aird	51,067. 7.10
J.T. Firbank	44,781.19. 7
H. Lovatt	52,750.14. 8
Mowlem & Co.	56,165. 0. 0
Pauling & Co.	52,789.17. 0
Perry & Co.	59,771. 0. 0
Pearson & Son	58,232. 0. 0
Relf & Son	48,489.11. 8

These prices should be compared with the costs of building a railway to main line standards. The best comparison is with the 1895 Portsmouth, Basingstoke and Godalming proposal which estimated the line between Basingstoke and Alton as costing £450,000. The Light Railways Act certainly allowed for significant cost savings to be made, although the specification was necessarily significantly lower.

On 20th July 1898 the LSWR announced that the lowest tender, that submitted by J.T. Firbank had been accepted. Messrs Firbank were no strangers to railway construction having been in such work since 1886. The company worked extensively for the Metropolitan Railway and had successfully tendered for previous LSWR contracts. Those in Hampshire included the Christchurch to Brockenhurst and Hurstbourne to Fullerton lines. At the time of tendering Firbanks were actively engaged in constructing the Canfield Gardens to Marylebone section of the Great Central Railway's London extension.

As was customary at the time a ceremony of 'cutting the first sod' was held. This was enacted on 22nd July 1898 in a field known as Sixteen Acres, Basingstoke. Subsequently it would be partly occupied by the engineering firm Thornycroft's. As befitting the first railway authorised under the Light Railway Act, the LSWR directors made much of the affair with a special train bringing invited guests from London to Basingstoke, one of whom was the former General Manager Sir Charles Scotter.

Punctually at 1.00 p.m. the Chairman of the Board, Wyndham Portal, invited the chairman of the Board of Trade, Charles T. Ritchie, to proceed, a silver spade with mahogany handle being used to move the first earth. Other dignitaries followed suit after which the spade was presented to Mr. Ritchie as a momento. The party then adjourned for a formal luncheon at which Ritchie outlined details of progress made under the Act. He stated that some 121 applications had been received representing 1,305 miles of new railway at a capital cost of £7½ million. On this basis he pointed to the success the legislation was likely to achieve.

It is worth mentioning that by 1922 a more realistic view was taken when the Ministry of Transport reported that whilst "a large number of Light Railway Orders were made, the actual result in mileage of lines constructed was disproportionate and disappointing."

Cliddesden Station: the woodcut illustration in the June 8th 1901 edition of the Hants and Berks Gazette. The photograph from whence it came appears on page 42.

The Post Office and shop at Herriard in 1920 with the main road from Basingstoke to Alton in the foreground. Unlike today, the road traffic was limited, environmentally friendly and sustainable.

Mr. Hooker of Hooker's Place, Bentworth who is thought to have acted as the Bentworth village carrier. Local carriers fulfilled a vital role as feeders to and from the railway system and in the horse drawn era complemented the railway. This view was taken outside the Swan Hotel, Alton. In 1905 G.P. Grove was the carrier from Alton to Basingstoke, leaving the Swan Hotel, Alton every Wednesday and Saturday.

Bentworth in a more tranquil age when horse traffic predominated c.1900.

CHAPTER 3
THE MAKING OF THE LINE

Construction near Winslade involving stationary steam engine power. The sharp curvature shown is no doubt temporary and may be connected with the construction of a bridge judging from the number of bricks lying around. Regretfully it is not possible to identify the buildings seen on the opposite side of the boundary fence although these may well be farm accommodation and would seem to be almost new. The rural nature of the terrain will be noted.
(Railway Magazine)

Unlike so many railways the construction of the Basingstoke to Alton branch is relatively well documented and accordingly worth recounting in detail.

The LSWR's solicitors sent out the requisite compulsory purchase orders in June 1898 and soon the process of land requisition commenced. The first purchase was completed on 21st September 1898 when L.W. Hitchcock of Alton sold 4 acres for £650. The process continued until June 1900 by which time all the required land had been purchased. Where agricultural land was involved the railway company was forced to compensate the landowner or tenant farmer for loss of crops and/or timber.

In all 121 acres were acquired from eighteen different vendors; the four largest landowners relinquishing 87½ acres. The importance of the large estate owners in the area can be clearly seen. The largest of these was the 5,800 acre Herriard Estate which sold 43¾ acres. Unable to agree terms, an independent arbitrator was appointed, the total sum eventually being paid to Mr. F.M.E. Jervoise amounting to £5,200.19/- or nearly £120 an acre. Included in this were sums for temporary loss and damage, timber and gravel. Land prices were higher in Alton and Basingstoke, Winchester College receiving £230 per acre and Lord Bolton £214 per acre. At the other end of the scale the Earl of Portsmouth settled for £62/10/- an acre for his holdings at Cliddesden. Even this was more than double the generally recognised sum of £30 an acre for agricultural land then considered applicable in north Hampshire. Scotter's pious hope that landowners would dispose of their land at advantageous rates failed totally with the result that land costs were double those estimated.

Meanwhile Firbanks had been busy extending their sphere of operation, winning several contracts with the LSWR. In August 1900 they were appointed the main contractor for the widening of Basingstoke station and the construction of a new down side goods yard to the west of the station. Other contracts won were the remodelling of Salisbury station and the building of the Amesbury branch (another light railway).

Firbanks started work on the Basingstoke to Alton line in September 1898. A siding connection at Basingstoke was requested by them and this was agreed to. Representing the contractor on site were his resident engineer Alexander Donaldson assisted by John Fraser as engineer and John Scott as agent. To create the necessary earthworks a variety of plant was used including a 'steam navvy' as well as stationary engines and hand carts. A number of small steam locomotives were also used:

Name	Arrangement	Maker
"Fox"	0-6-0 ST	Fox Walker, Bristol
"Henry Appleby"	0-6-0 ST	Hunslet, Leeds
"Portsmouth"	0-6-0 ST	" "
"Cliftonville"	0-6-0 ST	" "
"Newport"	0-4-0 ST	" "
"Walsall"	0-4-0 ST	" "
"Brockenhurst"	0-6-0 ST	Hawthorn Leslie
"Weaste"	0-4-0 ST	Hudswell Clark

It is probable that they were interchanged with the Basingstoke widening contract as work on that developed. The firm also established a small forge on

The construction of an underbridge near Winslade clearly showing the temporary decking. Passing over the top is an unidentified contractors' locomotive.
(Railway Magazine)

land close to Butts Junction, Alton.

The earthworks were relatively easy to work with few unexpected difficulties. The open downland meant that solid chalk was the usual subsoil although one cutting between Cliddesden and Herriard was of clay. This caused a certain amount of difficulty due to previous heavy rain.

Despite the purchase of enough land for a double line of rails, most of the earthworks were constructed to take a single track. The exceptions were the cuttings at either end and two embankments at Swallick and Winslade, which were widened to take double track. However to keep costs to the minimum large scale engineering features were avoided where possible, resulting in a line that was tortuous with numerous sharp bends and heavy gradients. It was whilst excavation was being carried out that a 7th century Anglo-Saxon burial ground was unearthed at Basingstoke. Apart from human remains, several artefacts of the period were uncovered including drinking vessels and bone shanks for buttons. Parts of another skeleton were subsequently discovered near the Butts, Alton.

Despite the actual works progressing smoothly, behind the scenes discontent was apparent. The first signs of this showed on 29th September 1898 when Cliddesden Parish Council wrote to the LSWR objecting to the proposed station site on the

One of Firbank's locomotives used in the construction of the line.
(Railway Magazine)

Basingstoke Road. The LSWR refused to reconsider the matter and on 2nd November 1898 the Parish Council wrote again with copies of the resolutions passed at the Council Meeting the day before. Again the request was refused. In desperation the Council appealed to the Light Railway Commissioners who had already offered to mediate if required. As a result an enquiry was held on 12th January 1899 at which a sensible compromise was reached. The station could not be placed where the council had wished due to the steep gradient of the line. However it was also accepted that the original site proposed by the LSWR was inconvenient. Accordingly a new position was suggested with the LSWR building and maintaining a new access road.

This uncompromising and superior attitude of the LSWR did little to endear it to the other local Councils. The railway company's attempts to operate eight out of the ten level crossings along the route as 'open crossings' met with fierce opposition from Alton and Basingstoke RDC's as well as the County Council. All objected to the potential danger for "persons, cattle and sheep using the roads." An objection was also received from a Mr. Bird of the Cliddesden Board School over the station level crossing. To investigate these issues the Light Railway Commissioners became involved again holding a hearing at Basingstoke on 28th June 1899. After hearing evidence as to the insignificance of the roads in question they agreed that five could be dealt with as requested by the company, the other five to be gated. Three could be overseen from the adjacent stations whilst the other two were to be provided with gatemen to be housed in purpose built cottages.

Regular reports from Galbraith show that construction work progressed steadily through late 1898, all through 1899 and into the middle of 1900. The influx of navvies caused local labour shortages with many agricultural labourers attracted by the relatively good wages. With the concurrent building of the Meon Valley line, Alton in particular had a significant number of temporary residents. Consequently accommodation was at a premium and in an attempt to meet the temporary demand a wooden hut was erected in Rack Close.

Not all the local population were against the influx, for despite the navvies' reputation as being tough, drunken and illiterate, their money was good and the various inns did a roaring trade; it was just as well Alton was a brewing town! In contrast regular bible classes were held by Miss Isabella Lewis of Queens Road at the new lodging 'house', appropriately known as 'The

An intermediate stage in the excavation of a chalk cutting with the steam navvy having scraped away to the basic depth after which human effort will be used to finish the slope. A variety of contractor's open wagons can also be discerned.

(Railway Magazine)

Near Whitedown, Alton the railway encountered one of its deepest cuttings, again through chalk, which was partly excavated by mechanical means. Note the timber shoring used at first and also the bowler hatted foreman. The virgin chalk revealed by the excavation at first gives the impression that the engine is almost superimposed on the picture although closer examination shows it to be standing on some temporary trackwork. *(Railway Magazine)*

Navvies' Hut'. A navvy mission was also established to raise funds for the workers' families with the not inconsiderable sum of £140 being secured. Assistance came from Wyndham Portal of the LSWR who held a garden party at his house whilst the Vicar of Basingstoke, the Rev. Dr. Cooper Smith also took a keen interest in the work.

It was at this time that the line's construction led to its first fatality. On 12th April 1899 a labourer, Arthur Vass was killed by a fall of chalk which crushed his head against a wagon buffer. He had only been employed a week being engaged by Thomas Birt, a sub-contractor. He was a local man, in his early 30s and a native of Preston Candover. At the inquest it was reported that it was customary when a fall of earth took place to get to the other side of the wagons as a form of protection but apparently Vass did not hear the warning shout. A verdict of accidental death was recorded.

Whilst the earthworks were being carried out the LSWR Engineering Committee considered additional details. In December 1898 it was decided to fence the line with iron post and wire rather than quickthorn. This was seen as being less expensive and more durable than the hedging which would not grow in certain places. Water supplies at the stations needed to be provided and it was agreed to sink wells at all three locations. It was recognised that housing would be needed for the staff and in April 1899 Galbraith suggested the provision of a house for each Station Master and a terrace of four cottages at every station for the staff. In May the arrangement of the stations and sidings was approved by the Traffic Committee.

June 1899 saw the works sufficiently advanced for a temporary junction to be put into place at Butts Junction, Alton with the Mid Hants line. At the same time a request was received, and agreed to, for a siding to the Steam Carriage and Wagon Co. Ltd premises at Basingstoke. Subsequently the business was acquired by Messrs Thornycroft & Co. Permanent way materials were now being ordered. It was estimated 26,000 sleepers would be required for which tenders were invited. The Southampton firm of Burt, Boulton and Haywood were successful in their bid for 'adzed' sleepers at 4/7½d each delivered to Basingstoke or 4/8½d each delivered to Alton.

By March 1900 the earthworks had been completed with the permanent way laid from Basingstoke to Herriard and from Alton to Bentworth and Lasham. The Steam Carriage and Wagon Co. siding was also completed and it was in use by them. The 2½ mile section between Herriard and Bentworth and Lasham was laid in the spring and all was complete by June. The first trip over the complete distance of the new line was arranged by the contractor's agent, John Scott

A remarkably clear view of a steam navvy at work on the light railway which was clearly a vast improvement on muscle power alone. Beside the metal and girders of the actual machine the operator can be seen standing on a small platform to the left of the jib whilst at the rear of the machine the corrugated and wooden sheeting afforded protection to the man responsible for maintaining a head of steam. The steam navvy would normally run on rails and it would appear its position alongside the temporary trackwork is to allow for excavation at this point - regretfully the location is not given. It is possible to discern at least 13 men in the photograph. In the background may well be a contractor's site area; these are not referred to in official records.

Building up an embankment with the aid of end tipping wagons. Large amounts of excavation were not needed to afford sufficient spoil for the embankments; instead quantities from the various cuttings were taken to Basingstoke to assist in the foundations for the extended yards.
(Railway Magazine)

on 29th July 1900 and this carried a number of dignitaries as well as those who had been fund raising for the Navvies Mission. On the 1st August the LSWR Engineering Committee heard that the "contractor's work is practically finished." It was also reported that the Basingstoke end of the line was being used for the movement of chalk for the expansion of Basingstoke station and yard.

Final touches were now being carried out. The signalling was ready by the end of October, whilst two out of the three wells had been sunk. That at Herriard was proving problematical and needed to be deeper than expected. Apart from this, all looked set for the line to be opened by the end of 1900.

It was in October there exists the first mention of a siding for the new hospital at Alton for wounded soldiers from the Boer War. The railway estimated the cost of the facility at £11,640 which was approved by the Hospital Management Committee, then operating

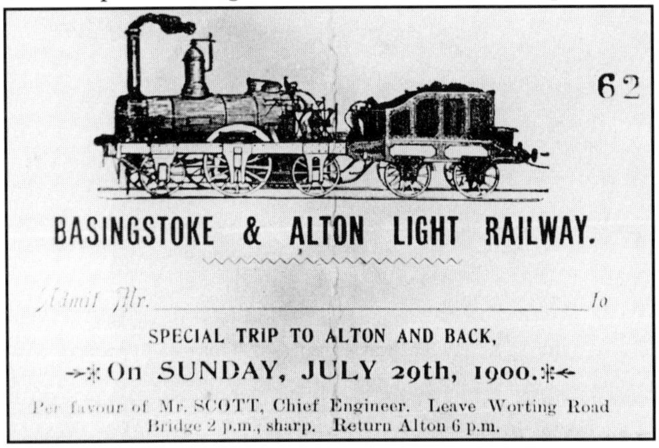

Invitation to participate in the first trip over the light railway on 29 July 1900 and destined to be almost a year before public services would commence. The drawing of the locomotive is somewhat misleading!
(Hampshire County Museum Service)

under the name of the "Absent Minded Beggar Fund." It would also appear that Basingstoke RDC, concerned about the provision of level crossings, inspected the line at about this time. Thus began the long and acrimonious correspondence between the Council and the railway company concerning Herriard station level crossing across Bagworth Lane. It lasted nine months from 1st October 1900 until 25th June 1901 and was totally inconclusive. Alton RDC and the County Council were drawn into the fracas both of whom sided with their Basingstoke colleagues. It appears no action was taken, the matter fizzling out after the train service commenced. Nevertheless much bad feeling had been created which did not bode well for the future.

The attitude of the local authorities (particularly Basingstoke RDC) with their continual criticisms over level crossings was irrational considering the limited road traffic of the area. It can only be concluded that the councils failed totally to understand the concept and method of operation applicable to light railways. Even the Inspecting Officer of the Board of Trade, when examining the line just prior to opening commented that "the action of the road authorities..... does not appear to coincide with the best interests of the public."

By the beginning of 1901 the time allowed for the completion of the line had elapsed and questions were being raised in the local press as to when the railway would open for traffic. Local frustration led to Basingstoke's MP A.F. Jeffreys giving notice to the House of Commons that he was going to ask the President of the Board of Trade when the line was scheduled for opening. Panic now set in. On 13th March 1901 the Board of Trade *hand delivered* a letter to the LSWR asking for "observations" to assist the President to frame his reply, and could the answer be given *this day*. Charles Owens, the General Manager

Some of the visitors who enjoyed the first trial trip in July 1900, seen arranged in a converted contractor's wagon - this was by no means a unique form of accommodation for trial trips on railways. Among those present were engineering and local officials as well as those who had been involved in raising funds for the local navvy mission. It is not clear how many vehicles the train contained At the head is 0-6-0 saddle tank 'Henry Appleby', built by the Hunslet Engine company in 1870 and previously used on a number of other Firbank contracts including construction of the Ross to Monmouth, Coton Park Branch, Derby to Eggington Junction, Rickmansworth to Chesham, Marylebone to St. Johns Wood, and Clarbeston Road to Latterston contracts. *(Hampshire County Museum Service)*

explained that the railway company had "encountered serious difficulties in providing terminal accommodation at Basingstoke due to alterations and improvements there." On 14th March Mr Jeffreys tabled a notice of his question to the President. As he said the "railway is complete when will it be opened?" On the same day, in a written note he was asked to postpone the question and the day after the President spoke to the MP. The answer must have satisfied him for Jeffreys informed him that he would "probably not put the question in the circumstances." It also satisfied the Board of Trade for there is no record of them invoking the £20 per day penalty for non-completion.

The correspondence definitely spurred the LSWR management into action. On 18th March 1901 they wrote to the Board of Trade requesting permission to "run as many mixed trains over the line as may be required." The Board concurred for in a letter dated 30th April 1901 permission to run three mixed trains was given, subject to satisfactory inspection.

Notice was then given to the Board of Trade on 1st May 1901 that the new double line between Alton and Butts Junction on the Mid Hants line over which the light railway trains would run would be ready for passengers within a month. Presumably further correspondence followed for less than one week later on 7th May 1901, Major Pringle inspected the new works.

Matters were now reaching a conclusion. On 17th May the LSWR informed the Board of Trade that the Light Railway was complete and requested a formal inspection. This was again performed by Major Pringle who carried it out on 22nd May. He found only a few areas of complaint, but nevertheless sanctioned the line safe for public traffic. His report is reproduced below and, as is usual with a Board of Trade inspection, it is highly informative:-

> "I have the honour to report for the information of the Board of Trade, that in compliance with the instruction contained in your Minute of the 18th int. I made an inspection yesterday (22nd May 1901) of the Basingstoke and Alton Light Railway, of the L. & S.W. Railway Company. This was the first railway authorised under the Light Railway Act of 1896.
>
> Of the two railways authorised by the Order of 1897 as amended by the order of 1900, that portion of Railway No. 1 from its commencement to its junction with Railway No. 2 has not been constructed. The line therefore commences by a single junction with the company's main line at Basingstoke and terminates on the Mid Hants

Butts Junction, Alton, c.1900 showing construction work in progress on the Meon Valley line. The light railway can be seen to the right curving in sharply to join the Mid. Hants line. This was not the final layout and shows the temporary connection put in to assist the contractor in the construction phase. *(D.H. Foster-Smith).*

branch at Butts Junction between Alton and Medstead stations.

The length of railway constructed is 12 miles 35.38 chains. Land has been purchased for a double line, but the railway is single throughout, with the exception of a loop at Herriard Station, and a length of double line at Butts Junction. The total length of double line is about 16 chains.

The gauge is 4ft $8^1/_2$ in. The space between double lines is 6 feet, and there is a minimum clearance of 2ft 4in between the sides of carriages and fixed works at a level of 3 feet above rails. The width of formation is from 17 to 18 feet.

The steepest gradient has an inclination of 1 in 50, and the longest such gradient has a length of $83^1/_2$ chains. The sharpest curve has a radius of 9 chains. The highest bank and deepest cutting have a height and depth respectively of 39 and 34 feet. The soil encountered has been chiefly chalk, with some clay.

The permanent way consists of steel double head rails varying in weight from 75 to 82lbs per yard in 24ft lengths. The rails are secured to cast iron chairs weighing $39^1/_2$ lbs by oak keys, and the chairs are fastened to the sleepers by three trenails and spikes. The sleepers are of creosoted fir measuring 9 feet by 10in by 5in, and are laid on gravel ballast, which has a stated depth of 12 inches (except in the case of a chalk cutting).

The railway is fenced on both sides, partly with post and rail and partly with wire fencing.

In clay cuttings 6 inch land drain pipes have been laid down for drainage purposes.

Bridging - Under. There are seven segmental arched underbridges each of a single span from 12 to 30 feet. In one case the railway is carried over a span of 12 feet on wrought iron troughing. This gave a moderate deflection under test load and has sufficient theoretical strength.

Over. There are two segmental arched overbridges each with a span of 15 feet.

Culverts. There is one culvert of 4 feet width. All the masonry work in these bridges is of brickwork and appears to have been well and solidly constructed and shows no signs of yielding.
Level crossings. There are 41 of these in all, of which 9 are footpath crossings. 20 are occupation crossings, and 11 (1 is a new road) are parish or public road crossings. All the occupation crossings are provided with gates opening away from the railway.
Of the public crossings, 8 are provided with gates, which normally fence in the railway, and as at present erected close entirely across the road, so as to fence in the road, when they are open for the passage of a train, the remaining 5 are provided with cattle guards.
From the point of view of safety, to both railway and road traffic, the proposed arrangements (when in all respects completed) at the public crossings meet with all the requirements of the Order, and are in my opinion the most suitable that could be devised.
Stations and signalling arrangements - Certain additions have been made at Basingstoke in 'B' and 'C' signal boxes for temporary working of the light railway. They are not very satisfactory, but as a temporary arrangement, in view of the extensive alterations contemplated at Basingstoke station, may I think, be accepted. The down bay line is at present proposed to be used both for arrival and departure of trains on the Light Railway.
There will be tablet sections on the Light Railway the proposed arrangements at the following points:

'B' box Basingstoke
'C' box Basingstoke
Herriard Station
Butts Junction

The Company have not yet reported the alterations affected at B and C Boxes. These I inspected but I will reserve my report until the Company's letter on the subject is received. There is a siding connection, for Thornycroft's works, on the single line close to Basingstoke, which is controlled by a ground frame released by the train tablet. The frame contain two levers. The points are fitted with a one way lock for the facing direction.
Cliddesden station - There is a single line and platform with a small shelter here, also a siding, the arrangements for which are identical with those described for Thornycroft's works. There are no signals. The platforms at all stations are 3 feet height, and they are of sufficient length.
Herriard Station - This is a tablet station and crossing place for trains. There is a platform on either side of the loop and a small shelter on one of the platforms. A siding connected with the east loop line is worked from a ground frame containing a single lever. The frame is released by the Herriard - Alton tablet. Home signals are provided at each end of the loop, and these detect the points, a rod being provided between the detector and signal. The points are worked with a self acting weighted point box, and are fitted with a 25 foot locking bar and special cross rod lock.
Lasham station - The arrangements are similar in all respects to those at Cliddesden, and there is the same siding.
Hospital siding near Alton. The Company agreed to take up the points, as the siding is not at present required, and to ask for the inspection of the siding when necessary. The arrangements at Butts Junction, the other terminus of the Light railway have been described in my report on the widening of the Company's line between Alton and Butts Junction. The light railway forms a double junction with the Winchester line and is provided with distant and home signals in both directions.
I make the following requirements:-
'Basingstoke 'C' Box - No. 20 lever to lock No. 17 both ways.
No. 20 (Alton) overbridge. The telegraph insulators fixed outside the arch were foul of the clear space necessary for a passing coach and require to be removed or refixed.
The points for the Hospital siding near Alton to be removed.
The erection of speed posts at Nos. 13, 18, 36 (Herriard) and other level crossings. These were not completed at the time of the inspection.
The position of some of the noticeboards on the roads at level crossings was not in accordance with the amending order.
Others require to be erected.'
Exception was taken on behalf of the Basingstoke Rural District Council to the encroachment on the public highway of the gates at No. 23 (Herriard) level crossing. The width of the highway appears to be about 45 feet, the gates are 15 feet wide. The metalling on the highway is about 12 feet, except at the crossing where it is 15 feet. The encroachment is clear, and the District Surveyor, who attended, said he would inform his council of the views held by the Board of Trade that they could not compel the Company to close their gates across the road, during the passage of train on the Light Railway, and take their instructions on the point of issue. The Company expressed their willingness to remove the encroachment if so desired and pointed out that the gates would not in this case close across any portion of the metalling.
No. 3 (Lasham) level crossing. The Alton Rural District Council objected on the score of encroachment on the highway. Whilst not proposing to admit the encroachment, the Engineer of the Company agreed to meet the views of the Council so far as to give a clear width of 24 feet on the east side of and on the crossing itself.
No. 21 (Lasham) - This is a County Council road and I understand they object to the crossing as an encroachment. The engineer of the Company said he would communicate with the council and remove the gates if they required it.
The action of the road authorities in all cases does not appear to coincide with the best interests of the public. The traffic in all cases on the public roads is small, and the actual width of metalled roadway provided by the Company is greater than what formally existed.
I observed in the case of two arched underbridges No. 15 and No. 6 (Cliddesden) that the dimensions in the Order had not been adhered to. No objection however appears to have been made by the road authorities.
Subject to the completion of the requirements made in this report and without prejudice to any of the points of law involved I can recommend the Board of Trade to certify the light railway as fit for passenger traffic.
I have the honour etc."
Signed: Maj. Pringle

The 12 ½ miles of railway had taken three years to complete as opposed to the twelve months the 'Railway World' had suggested in 1898. Seventeen years after the first proposal to build a railway between the two towns, everything was at last complete. England's first railway authorised by the 1896 Light Railway Act was about to open for business.

Part of the extensive down side yard at Basingstoke which was extended at the same time as the light railway was being built. The extent of the earthworks is clearly seen, much of the soil coming from the light railway whilst it was being built, thus delaying its opening. This June 1925 view shows a motley collection of vans for perishable traffic including a number of ex-LNWR vehicles. In the centre is Basingstoke West Signal box which controlled access to the light railway from the north.

CHAPTER 4
A RURAL RAILWAY – THE LINE DESCRIBED

Trains for Alton officially ran in the 'up' direction. Confusingly, they left from the 'down' bay platform at the west end of Basingstoke station and ran on an independent line sandwiched between the down main lines and the down goods yard. Safe operation of the branch was ensured by the use of 'Tyer's tablet' working, this vital component being collected by the crew of the locomotive on passing Basingstoke West signal box. This meant that unlike some other main line stations where there could be a simultaneous departure of main line and branch services, there was little opportunity for the crews to race each other. The initial departure involved a gentle climb, starting at 1 in 333, increasing to 1 in 262 and then 1 in 218.

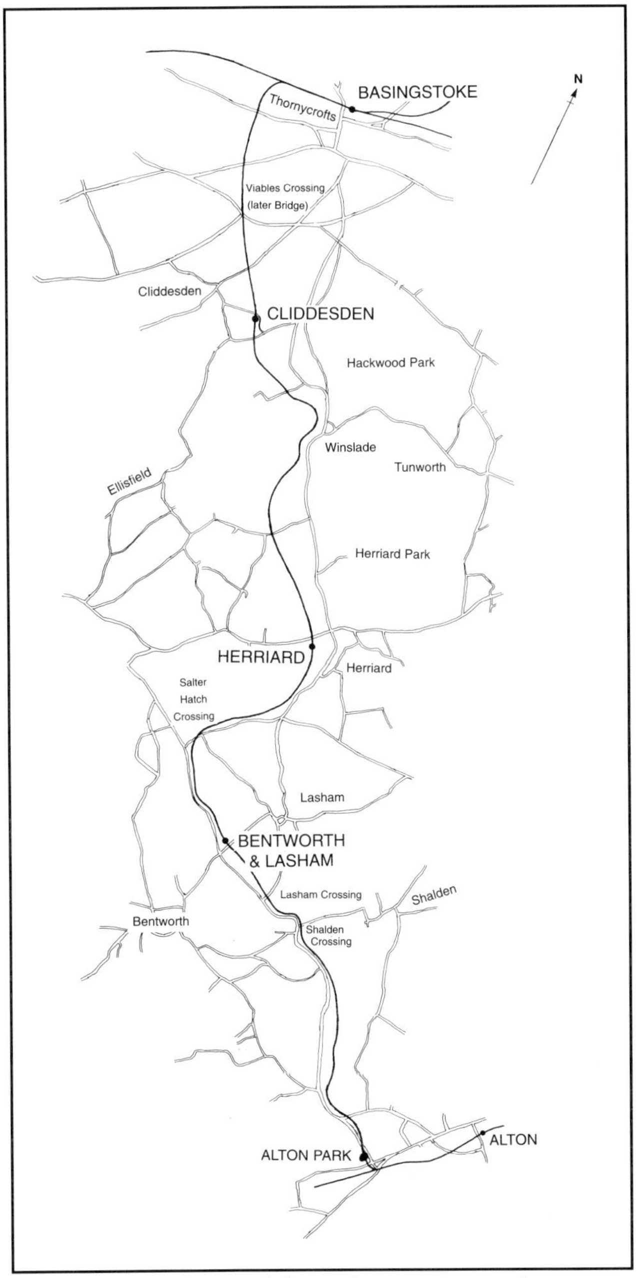

The railway and the settlements it served.

Branch and main lines ran parallel for about ½ mile before the light railway turned sharply south on a 17 chain radius curve whilst descending a 1 in 82 gradient through a shallow cutting. Part way round the curve and at a point where the cutting opened out, existed a level crossing. This was used to gain access to the Basingstoke Water Works, built in 1906 which was located on the west side of the line in an area known as West Ham. The keys to the crossing were held by Basingstoke UDC who used it on average twice daily. Plans have been found to suggest a siding was considered to serve the water works although this may have been solely to assist in its construction. It was never built, although, as referred to later, a siding for British Petroleum was provided in a similar position on the other side of the branch in later years. Close by the water works and opened at the same time were the public swimming baths. In 1906, when this facility was provided, the railway was asked to plant a line of trees along its boundary so as to furnish some privacy to the users of the amenity. At this point the line was still curving atop a shallow embankment. Underneath was a 4ft. brick arch through which flowed the river Loddon, little more than a stream at this point. King George's playing fields were to the left immediately before a brick underbridge which allowed the railway to cross Deep Lane. The line now straightened and levelled out passing Thornycroft's factory to the left and open ground on the right. In 1919 this was purchased by Thornycroft's and converted into a sports field, football ground and tennis court. A short rise of 1 in 50 afterwards led to Worting Road overbridge. Constructed of brick, it lay at right angles to the railway and afforded a maximum clearance of 14ft. 5in. from rail to crown. Around this point was a short descent of 1 in 264 whilst the line threaded a shallow cutting, banded by trees to the west and open land to the east. The latter was used as a cemetery from 1932 onwards.

The railway had now left Basingstoke behind and began its climb out of the Loddon valley on to the Hampshire downs. It now passed a pair of cottages whilst the embankment increased in height at the approach of the Winchester Road underbridge, this being the largest bridge on or over the branch with a span of 30ft. Situated on a slight skew the bridge had a maximum clearance of 16ft. 4in. Soon afterwards came Viables crossing. This location changed significantly when the Basingstoke-by-pass was built between 1930 and 1932. Crossing the railway on a ferro-concrete overbridge, the Southern Railway stipulated that it was to be sufficiently wide to allow for

An early view of Thornycrofts from the Worting Road complete with haystack. The light railway is just visible in the middle distance running behind the buildings.

A wet day in the works yard with a number of solid tyred windscreen less road vehicles ready for despatch. All were probably going for military use. The siding accommodation was extended considerably within the works over the years.

Thornycroft "J" lorry chassis destined for use on the LSWR and lined up for the camera shortly before delivery. From the writing on the chassis they were limited to a top speed of 12 mph.

A final view of the company's works showing vehicles destined for military use. The factory expansion is clearly seen in the middle distance whilst at the back of the yard the line of the light railway can just be discerned.

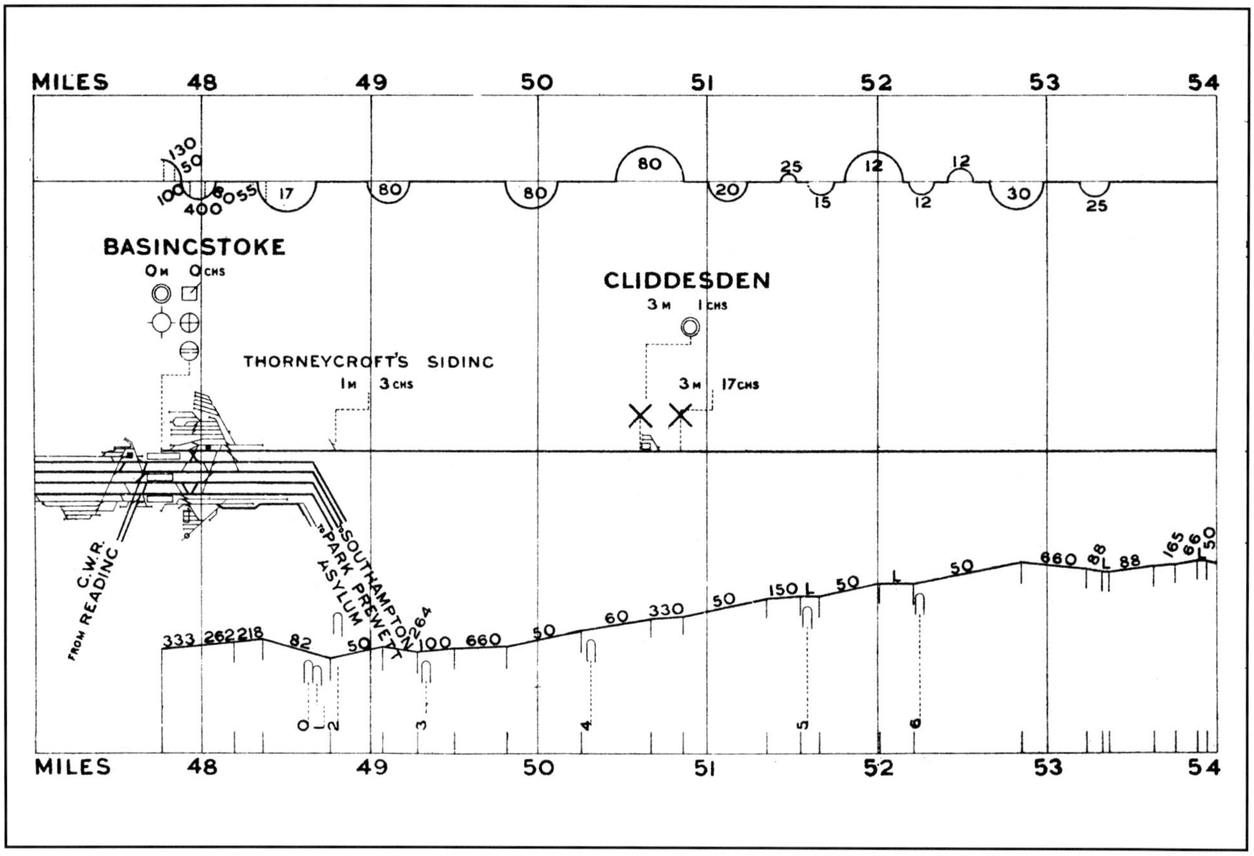

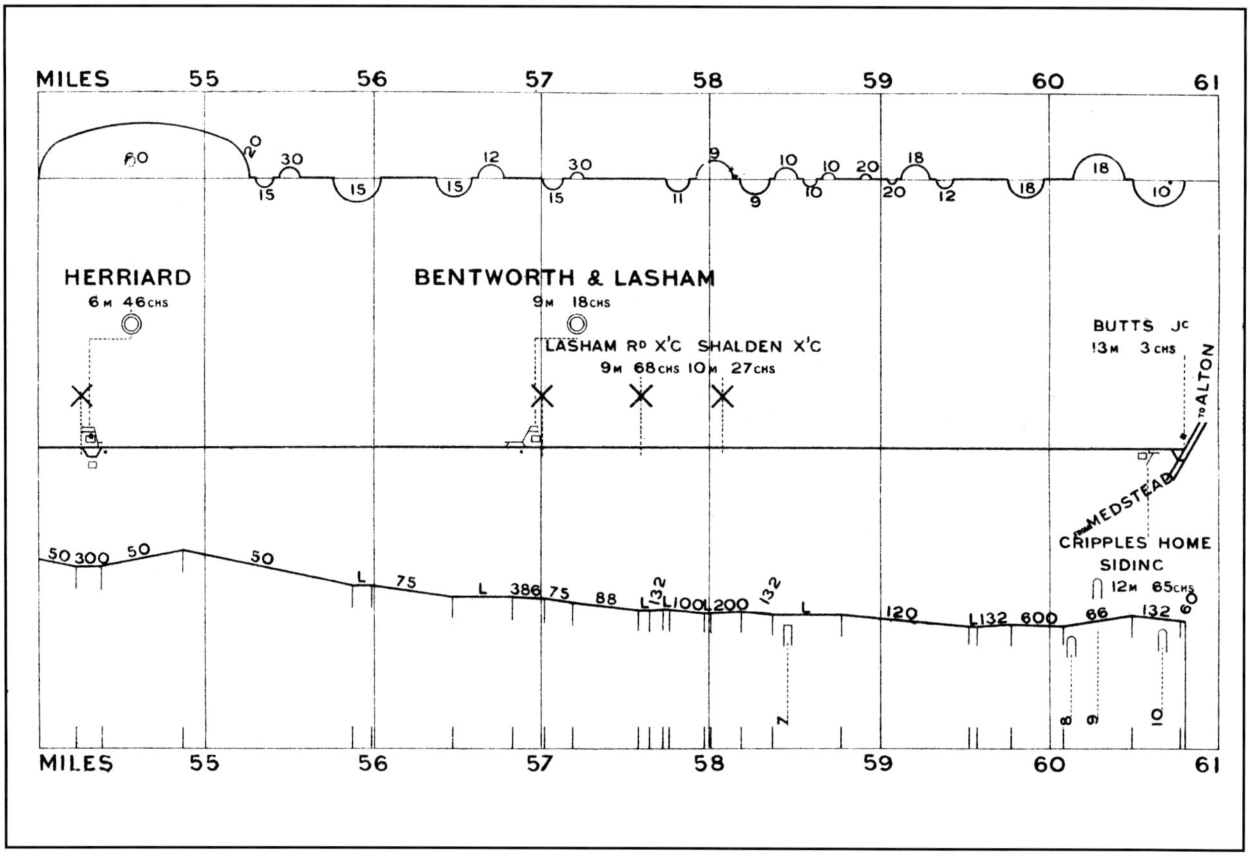

Mileage, curvature and gradient profile of the line.

a double line of rails, a somewhat surprising demand considering the traffic the branch was carrying.

The route was now straight, climbing steadily, initially at 1 in 100 but with subsequent sections of 1 in 50 and 1 in 60. For trains heading towards Basingstoke this long descent provided a welcome, if brief, opportunity for a breather. Despite the descent, train speeds were never excessive with the regular 'clickety-click' giving a false impression of speed due to the short 24ft. lengths of rail originally used.

With the line still climbing came the Cliddesden Road underbridge, another skew arch, this time of 25ft. width and 17ft. 6in. height. From its lofty perch the traveller was afforded a grandstand view of Cliddesden to the right, its cottages a random mixture of tile and thatch, the epitome of a village. The embankment was quickly followed by a short cutting coinciding with the start of an eighty chain radius curve to the right. The line was now in open downland as it crossed a level crossing over the road which the LSWR were required to construct and immediately entered Cliddesden station. The platform and small yard were on the east side of the line. Opposite the platform were the four station cottages, the Station Master's house and the water tank on a tower powered by a wind pump. The whole was (and still is) an isolated community set apart from the village it purported to serve. It was just possible to discern St.Leonard's Church giving alighting passengers a clue as to where the village was located.

Leaving the station the line crossed Hackwood Lane by a level crossing and continued to climb at 1 in 50 before swinging on a 20 chain curve to the left bringing the track into close proximity with the main Basingstoke to Alton road. Across the fields from the embankment could be seen the trees of Hackwood Park, the home of the Earl of Portsmouth. The railway was now crossing his Estate, undulating for a short distance. A brick underbridge with a 12ft. span took the line over Swallick Road whilst to the right were the appropriately named Swallick cottages and farm.

The climb now resumed with a vengeance, again at 1 in 50. This was not assisted by right and left hand curves of 15 and 12 chains radius as the line followed the contours of the land skirting Buckshorn Copse on the hill to the right. The rise continued for almost $\frac{1}{2}$ mile and was notorious as the location which posed difficulties for the steam rail motor services in 1904. There was now an embankment and to the left on the main road could be seen the small settlement of Winslade. With a population of only 59 when the line opened, its church, St. Mary's could be clearly seen. The embankment continued to increase in height and passed over Winslade Bridge which allowed a lane to pass underneath. From the vantage point of the bridge it was possible to see the village cottages to the left as well as the small Congregational Chapel hard by the railway to the right.

The route continued to swing alternately right and left through dense woodland. The embankment continued through Cowleaze and Fryingdown Copses until the line emerged into a shallow cutting as the woodland was left behind. This part of Hampshire was renowned for its pheasant breeding and hunting. It is therefore not surprising that Lord (later Earl) Portsmouth asked for the reservation of the sporting rights over the land taken by the railway as early as December 1899. He also asked "for his game keepers to be given permission to walk on the line to warn off trespassers." This was given by the LSWR, but it clearly showed the priorities of one of the major

PERMANENT SPEED RESTRICTIONS—*continued.*
THE UNDERMENTIONED SPEED RESTRICTIONS MUST BE STRICTLY ADHERED TO.

Down Trains. Maximum speed per hour.	Points at or between which speed must be reduced.	Up Trains. Maximum speed per hour.
Miles.		Miles.
	BENTLEY AND BORDON LINE.	
10	When approaching and within 300 yards of the following Crossings:—	10

Name of Crossing.	Distance from Bentley Junction.
Blackness Road	1 mile 10 chains.
Binstead Road	2 miles 49 chains.
White Hill Road	4 miles 17 chains.

25	Over other portions of the Line	25
	BASINGSTOKE AND ALTON LINE.	
10	When passing over curves between $57\frac{1}{2}$ and $58\frac{1}{2}$ mile posts, between Lasham and Butt's Junction	10
10	When approaching and within 300 yards of the following Crossings:—	10

Name of Crossing.	Distance from London via Basingstoke.	Stations between
Viables	$49\frac{3}{4}$	Basingstoke and Cliddesden.
Bushey Warren	$53\frac{1}{4}$	Cliddesden and Herriard.
Grange Road ...	$53\frac{3}{4}$	Cliddesden and Herriard.
Herriard Common	55	Herriard and Lasham.
Salter Hatch ...	$55\frac{3}{4}$	Herriard and Lasham.

25	Over other portions of the Line	25
	NEW GUILDFORD LINE.	
20	29 MILE POST AND GUILDFORD YARD BOX	20
	BISLEY CAMP LINE.	
10	BROOKWOOD AND BISLEY	10
	KINGSTON AND THAMES VALLEY LINES.	
	FROM THE FLY-OVER BRIDGE TO TWICKENHAM STATION ...	15
	READING AND WINDSOR LINES.	
15	STAINES JUNCTION AND THAMES RIVER BRIDGE (Reading Line)	15
15	STAINES HIGH STREET STATION AND STAINES SOUTH JUNCTION	15

Permanent speed restrictions on the line.

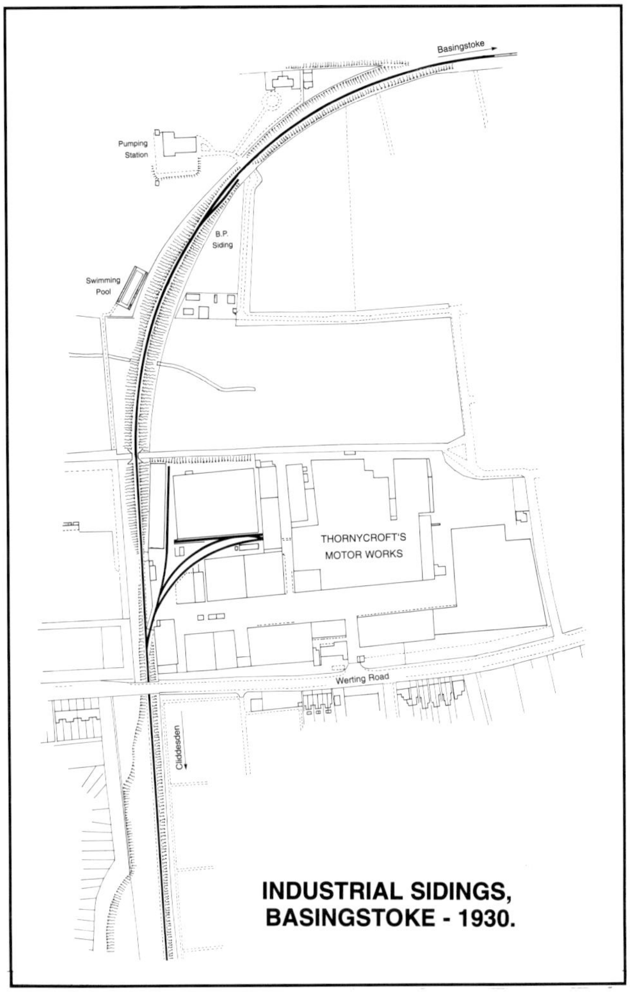

INDUSTRIAL SIDINGS, BASINGSTOKE - 1930.

estate owners. Nevertheless it was inevitable that an accident would occur; it happened around 1930 when two hounds were reported as killed by a train near this point.

An undulating section of line across open fields skirting Great Busheywarren Copse followed. This was the first relatively favourable section for 'up' trains in almost four miles. The line was now almost straight as it crossed two level crossings in quick succession. The second of these crossed Scratchface Lane which led to Manor Farm, Herriard and was the scene of an accident in 1908 which is described later. Past the farm the lane led to the main road and the entrance to Herriard Park, the home of the Jervoise family.

A shallow cutting and low embankment on a gentle 60 chain radius curve down a gradient of 1 in 50 led to the controversial level crossing across Bagmore Lane. Across this, the line immediately entered Herriard station, the only passing place on the line up until 1916.

From the station it was possible to see the village school which was located close to the level crossing whilst nearby was Ellisfield House which served as the vicarage for many years. Being the only passing place, Herriard was the only station to be provided with fixed signals. As with Cliddesden there was a small yard to the east of the line behind which was the Station Master's house. The row of four cottages for the staff were isolated from this, being on the west side of the line fronting on to Bagmore Lane.

Leaving the station the line passed through a shallow cutting with the water tank high up on the left hand side. A right hand curve and the final ½ mile 1 in 50 climb took the line on to a low embankment, where Bull's Farm could be seen to the left adjacent to the main road. It was here that the summit of the line was reached at 596 feet above sea level. Save for a few minor exceptions it was downhill now for over six miles, the start of the descent being marked by a long section at 1 in 50, followed by a shorter section of 1 in 75.

The railway was now in open countryside and passed on the right a couple of old chalk pits. To the left was a minor road which paralleled the line as both skirted the open expanse of Lasham Hill to the left. The demands of World War II led to the building of an aerodrome here in 1943. This severed the main road necessitating it to be diverted along the route of the minor road and the trackbed of the by then abandoned railway.

A sharp left hand bend skirting some old gravel pits took the line over Salter Hatch crossing, the site of the derailment in the film "The Wrecker". The railway, still paralleling the minor road now took a succession of alternating left and right curves, some as sharp as 12 chains radius. Forever descending, more abandoned gravel pits could be seen to the right as the line clung to the hillside crossing open fields. Meanwhile the road had descended a steeper gradient to the bottom of a shallow valley.

Straightening out the railway made the final approach to Bentworth and Lasham station passing the wind pump and water tower supplying the station's needs to the left. Immediately afterwards on the same side was the goods loop and the sidings behind the single platform. As at the other stations there was a terrace of four staff cottages and a detached house for the Station Master. All were on the eastern side of the station behind the goods yard. Just as at Cliddesden, the railway community was quite isolated from the villages the station was supposed to serve.

Leaving the station, the railway immediately crossed the road from Bentworth to Lasham by a level crossing. The descent continued through a shallow cutting passed yet another gravel pit. The location of these adjacent to the line was not coincidental, for when discovered in the construction phase the material had been used to ballast the line. The railway now crossed another road from the village on the level at Lasham crossing and here was the scene of a near fatal accident in 1925.

To the left was the crossing keeper's cottage - the first of the two on the line. If the route had numerous curves up to this point, from now on it became serpentine in character as it followed a narrow well wooded dry valley between high hills. After a sharp right hand bend the line crossed a country road on the level. This was the site of the second gated crossing which was stipulated by the Light Railway Commissioners. The Shalden gatekeeper's cottage was

Lord Mayor Treloar's Hospital; the power station to the right was the source of considerable traffic.

Another early view of Lord Mayor Treloar's Hospital. The railway line in the background is the Mid Hants line; the signal protecting Butts Junction is just discernible above the tree in the middle of the photograph.

on the right, located on a triangle of land between the railway, the main road and the lane to the village of Shalden. From here the daughter of the crossing keeper would regularly walk into Alton for work at 6.30 a.m. - there being no early train.

Sharp reverse curves of nine and ten chain radius now followed as the route threaded the valley. To the right was the extensive woodland of Binney Copse which came down to the trackside. Running on a low embankment the railway ran on to bridge number seven, an underbridge serving an occupation road which led to the fields on the left. For many years this was one of the few remaining bridges in situ on the line. Now it has gone, swept away by road improvements.

Following a gently undulating course the railway alternated between embankment and cutting passing Hassocks Copse and the larger Thedden Copse to the right with Great Wood to the left. Here the line straightened briefly before a succession of sharp reverse curves. With the route approaching Alton the embankment increased in height passing the delightfully named Hungry Copse, before it crossed over a lane - now the main A339/A32 road into the town.

The final approach to Butts Junction was noted for its extensive earthworks. The embankment quickly gave way to a cutting of 34ft. depth, the deepest on the line. Curving to the right the line passed under the skew angle overbridge of Whitedown Lane - a notorious danger spot for motor traffic. As the cutting slowly gave way to embankment the formation curved to the left with the private platform and siding to Treloar's Hospital on the right. This, the final curve on the line was of 10 chains radius and took the railway over Shalden Park Bridge to Butts Junction with its protecting home signal.

The junction was a relatively complex one with three single lines of railway joining at the same place. Immediately before the junction all three routes doubled. The light railway joined the Mid Hants route first and this was immediately followed by the Meon Valley line to the right. Unlike the Basingstoke branch, the Meon Valley joined the Mid-Hants at a shallow angle, clearly indicating it was built to main line standards. The 40 lever signal box was to the left immediately opposite the junction with the Meon Valley line. Crossing over the main Winchester road by a plate girder bridge the route into Alton was now of double track.

The whole section from Butts Junction to Farnham

Passing Great Wood on the sinuous approach to Alton.

Butts Junction, with the signal box to the left of the double track leading to Alton. The Basingstoke line swings in from the left and the Meon Valley branch is to the right. In the foreground is the Mid Hants branch from Winchester and Alresford.

had been doubled in 1901 in anticipation of the increased traffic to Fareham and Basingstoke. This led to a remodelling of Alton station where the existing 'down' platform was converted into an island to cater for the increased number of trains starting and terminating their journeys. Although trains from Basingstoke left in the 'down' direction, they entered Alton in the 'up'. Descending at 1 in 100 through a cutting the line passed the trailing connection to the brewery siding which was only accessible to Alton bound trains. Alton station was remodelled with three platform faces in 1903. Trains from Basingstoke could use either the 'up' main or 'up' loop platforms - the latter being on one side of the new island. Departures for Basingstoke normally left from the 'down' platform on the other side of the island. When the Basingstoke service ended and the line from Waterloo was electrified in 1937 the layout was remodelled whilst further changes took place after the Mid Hants line was closed in 1973.

Alton station was a busy location and passengers from the light railway would change trains for Fareham, Winchester, Guildford or London. Although only the latter service remains, the station is busier than ever and it does have the unique distinction of having two lines serving it being abandoned, only to be reinstated at a later date.

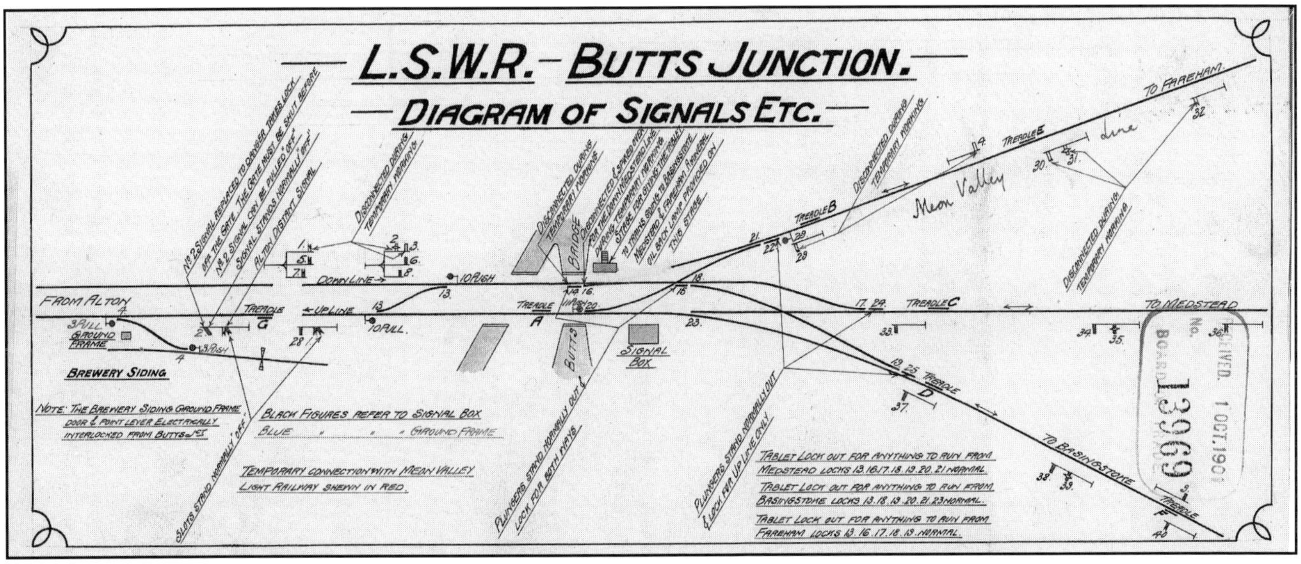

Signalling diagram for Butts Junction in 1901 before the Meon Valley line was brought into use.

Butts Junction in LSWR days with two of the three single line section tablets out and on display. At this time the box worked the double line to Alton and the three single lines to Herriard on the Light Railway, Medstead on the Mid Hants and Tisted on the Meon Valley line. *(Lens of Sutton).*

The areas known as 'The Butts' in Alton with the photographer standing in the main Winchester road. Just visible in the trees in the background is the girder bridge carrying the railway over the road close to the junction of the three lines.

"Q" class 0-6-0 No. 30535 works hard up the climb to Alton past Courage's brewery siding with a freight from the Mid Hants line, 25th October 1952. *(Denis Cullum)*

Drummond 'M7' class 0-4-4 tank engine at the head of a three coach Alton to Southampton service passing the brewery siding and approaching the former Butts Junction on 23rd September 1952. The Mid Hants and Meon Valley lines now ran independently into Alton as two single lines giving the impression of running 'wrong line'. *(Derek Clayton)*

C Bushnell.
CLIDDESDEN

LONDON AND SOUTH WESTERN RAILWAY.

INSTRUCTION No. 30, 1901.

Instructions to District Superintendents, Station Masters, Inspectors, Enginemen, Guards, Signalmen, and all others concerned, as to the

OPENING OF THE
BASINGSTOKE AND ALTON LIGHT RAILWAY
On SATURDAY, JUNE 1st, 1901,

FOR

PASSENGERS, PARCELS AND GOODS TRAFFIC.

A New Single Line, 12 miles 72 chains in length, has been constructed between Basingstoke "C" Box and Butt's Junction Signal Box. The gauge of this New Line is 4 feet 8½ inches. No Engine, Carriage, or Truck bringing a greater weight than fourteen tons upon the rails by any one pair of wheels must be run on this Railway.

The speed of Trains must not exceed 20 miles per hour at any time, or 10 miles per hour when passing over any curve the radius of which shall be less than nine chains.

For the present, Trains to and from the Alton Light Railway will travel between Basingstoke Station Down Bay Road and the "C" Signal Box on the Siding, next to the Down Main Line.

There are three intermediate Stations, viz. :—**Cliddesden, Herriard** and **Bentworth and Lasham.**

The Line will be worked under Tyer's Train Tablet System (No. 6 Instruments) as described in Instruction No. 17, 1898, which has been supplied to all concerned.

The Tablet Sections are as follows :—

1. Basingstoke Station Bay Road, ("B" Box) ... ... to Basingstoke "C" Signal Box.
2. Basingstoke "C" Signal Box ,, Herriard Station.
3. Herriard Station ... ... ,, Butt's Junction.

17

The official opening notice of the light railway issued to Charles Bushnell, Station Agent at Cliddesden.
(Hampshire Record Office.)

CHAPTER 5
THE EARLY YEARS

"This line will be opened for passengers and goods traffic on Saturday June 1st 1901" stated a small poster on Basingstoke station. The very fact that the notice was almost hidden from public view implies the lack of interest the LSWR were displaying in their latest venture. This view was articulated by Sir Charles Scotter three months later to the day when at the half yearly meeting of the company he said of the branch that it "is not likely to be of great benefit....but will at any rate be to some extent, a feeder to the main line." These were hardly words of encouragement.

Nevertheless aside from such reservations the railway was at last available for traffic. A graphic description of the line together with the opening day's events was admirably recorded in the columns of the 'Hampshire Herald and Alton Gazette,' complete with a certain amount of journalistic licence and sarcasm.

OPENING OF THE BASINGSTOKE & ALTON LIGHT RAILWAY

'Quietly and unostentatiously the train service on the new light railway between Basingstoke and Alton was inaugurated on Saturday. It is called a "light" railway because it was constructed under the provisions of the Light Railways Act, and was the first to be sanctioned in Hampshire, but as a matter of fact it is not a light railway, as the term is generally understood. It is of the same gauge as the rest of the LSWR system, and carries the ordinary rolling stock. But yet it is "light" in more ways than one - at any rate at present. It is light as regards the number of trains travelling over it, it is light as regards the loads of the trains, and it is extremely light as regards the station buildings. As we remarked last week, it was generally expected that the opening of the line, for which such a great future has been predicted, and which we were assured was to confer such vast blessings on the community, would be made the occasion of a great demonstration. That this idea seems to have been pretty widespread is evidenced by the fact that last week an enterprising firm of fireworks manufacturers, known the world over, wrote to the engineer of the line, saying that it was usual, on the occasion of the opening of a railway, to include a display of fireworks in the programme, and if such was contemplated on this occasion they would be happy to place their services at the disposal of the authorities. But no fireworks were wanted, for, as one of the company's officials remarked to the writer, "We finish before dark", in other words the last train for the day ran long before the sun had set. No, this "stupendous business" as the same official called the opening of the line, was not heralded with a fanfare of trumpets, cemented with a public banquet, or crowned with a display of fireworks. The train service began, it inaugurated "the trivial round, the common task" with a modesty thoroughly in keeping with the provision for the requirements of the public which the new line will afford. Trains leave Basingstoke at 9.15 a.m., 12.20 p.m. and 5.45 p.m., and return from Alton at 10.25 a.m., 1.55 p.m. and 7 p.m. so that if a person arrives at Basingstoke at 11.13 on business and fails to catch the 12.20 out, there will no other train available till a quarter to six. Rather a long wait, but then there is the consolation, as someone remarked, that one could walk back quicker.

Of course in making a railway there are lots of circumstances to be taken into account and heaps of difficulties to be surmounted which are not always appreciated by those who are not behind the scenes. The man in the street although he is supposed to know as much about some things as His Majesty's Ministers, is apt to look at things in a plain matter-of-fact manner and if he sees that a railway station is named after a certain village, he expects to find the village somewhere near the station. But such little details do not trouble the promoters of railways, who, if the station is somewhere in the parish named - the parish may be a dozen miles in area - are quite content to take the name which, with common people, is generally associated with the village. And so it has turned out on the Basingstoke and Alton Railway. The three intermediate stations are named Cliddesden, Herriard, and Bentworth and Lasham respectively. But the villages bearing these names are a long way from the stations and not a glimpse of them can be seen from the train. The village of Cliddesden is about a mile-and-a-half from the station, and in fact is almost as near to Basingstoke station, whilst Bentworth village is a couple of miles from the station with the double-barrelled name. Lasham, which takes second place in this double designation, is about half-a-mile from the station, and Herriard is a mile or more from its station. In fact the only houses near any of these stations are the cottages which the railway company have put up for their employees, with a "better-most" one in each place for the Station Master. The most imposing objects at each station are the windmills for the purpose of water supply. The station buildings proper are most primitive. They consist of little galvanised iron shanties divided into "booking offices" and "waiting rooms" in neither of which there would be room to swing a cat. There are no signal boxes, and with the exception of Herriard, no signals. The points are worked by hand levers on the ground, and the signals at Herriard are manipulated in like manner. One almost expected to find the old-fashioned half moon signals, but in this respect, at all events, modern ideas have prevailed. At the stations where such innovations are unknown, we get a reminiscence of the dim and distant past by seeing the Station-Master armed with a couple of flags. Speaking of the Station Master, we may mention that on the new line three old and well-tried servants of the company who have spent the best part of their lives at busy centres have been promoted to the charge of the three intermediate stations. These are Inspector Bushnell of Basingstoke, Inspector Hooper of Woking, and Inspector Pain of Southampton West, who become Station Masters at Cliddesden, Herriard and Bentworth respectively. Their duties will not be very heavy, but the best interests of the company and also of the public will be quite safe in their hands.

'And now we must say something about the initial ride over the line on Saturday morning. The train left Basingstoke at 9.15, and it consisted of an engine, two coaches, a guard's van, two trucks of coal, and two road boxes. The coaches and guard's van were all fresh-painted and upholstered. There were a good many passengers, but most of them were "free trippers". Amongst the latter, thanks to the courtesy of Mr. Barnes, the Station Master at Basingstoke, was a representative of this paper, the only Pressman, we may mention that travelled on the first train. The other free passengers were mostly officials, or subordinate employees of the Railway Company. There was quite a batch of the latter, some merely going for the sake of a ride, while others had to be dropped off at the different stations to take up their duties there. With such a little army of railwaymen 'on board', one had the consoling reflection that if the engine should fail there would be plenty of hands to push the train along, or if any mischievous individuals had, with Boer like tactics, thrown a tree across the line, it would have been a small matter to have removed it. Fortunately neither of these occasions of service presented themselves. The paying passengers did not number many more than half-a-dozen, and these all went for the sake of novelty. Alderman Powell of Basingstoke, had the distinction of purchasing the first ticket.

'A score or so of people assembled on the platform at Basingstoke to witness the departure, but there was not the least excitement. Mr. Scott, the engineer of the line, rode on the engine, and in the train were several officials who made the journey for the purpose of checking the times, testing the points etc. The train started from the siding on the east side of the down platform, and after 'feeling our way' down by the side of the new goods yard we were soon taking the curve which sweeps round Messrs. Thornycroft's works, and so on under the bridge which spans the Salisbury road. From the huts, which are still standing near the bridge, a few men, women and children turned out to give us a parting cheer, whilst a photographer

Purportedly the first day of services at Cliddesden with '02' class 0-4-4 tank engine No. 203 seen at the head of a mixed train for Alton. The safety chains on the front of the buffer beam were even then gradually falling out of favour. Mr Bushnell, the station-master, is in the centre of the group. The raw chalk can been clearly seen opposite the platform. *(L.G.R.P.)*

planted his camera on the clearing and possibly got a picture of the train as it emerged from beneath the bridge, or as it proceeded onwards. Save for a few boys and men astride of a gate, or perched on top of a bank, we saw nothing more of humanity till we reached Cliddesden station, our first port of call. Here there was a wait of about five minutes, not on account of the number of passengers to be dealt with, for there was none to be set down and not more than three to be taken up, but in order to shunt a road box. As mentioned above, there were two road boxes on the train and two trucks of coal. The coal was for Herriard, and one of the road boxes was for Cliddesden. As Cliddesden was the first stopping place, the most natural thing in the world was to have put that particular road box behind the coal trucks on making up the train at Basingstoke. But this would have been too commonplace for such a "stupendous" occasion, and so the Cliddesden wagon was put in front of those for Herriard, and consequently the coal trucks had to be detached at the platform, then we ran down to the points, shunted the road box into a siding, and eventually backed to the platform and again picked up the coal trucks. At length we started again and ran through Winslade, with its little church nestling close to the railway - there will be no Sunday trains to disturb the worshippers - and skirting Farleigh and other small parishes, passed through the extensive estate of Mr. F.M.E. Jervoise, in the centre of which Herriard station is situate. This is a more pretentious station that either of the other two on the route, for although its "offices" are very little if any bigger, there is a double platform, with a relief line of rails. Here there was a longer wait than at Cliddesden. The trucks of coal had to be shunted into the goods yard, but of course the remaining road box was behind them on the train, so that there were more trips up and down the station. The coal, it may be mentioned was consigned by Messrs. Stephen Phillips & Co. of Basingstoke, and Mr. Councillor Phillips, the head of the firm, who was among the passengers, saw the trucks safely deposited in the goods yard. The shunting of these rather disconcerted a pony standing in the yard, and two ladies who were with the animal had their work cut out to hold it. Meanwhile most of the passengers had alighted, and were taking a stroll on the platform. They might have 'done' a historic building in the time had there been one within reasonable distance. At length we were once more on the move. Hitherto the line had been far from straight, but thence to Alton it is a veritable serpentine character. It is all ins and outs and ups and downs, and this may account for the plentiful supply of boards bearing the injunction, "speed not to exceed ten miles an hour". Either through want of time, or because of the desire to give these boards a spic and span appearance, some of them were not finished. On one the painters were doing the notice, and we noticed that the driver seemed to treat this like the signal which Nelson couldn't see with his blind eye, for the pace was appreciably quicker over that length. Level crossings abound all the way - there being several highways, numerous crossings from one field to another, and lots of footpaths. Where the crossing is from field to field there are gates at each side bearing the customary injunction "Shut this gate", but save in one or two cases there is no barrier whatever to the highways, although there are some "grids" which are supposed to act as a deterrent to cattle. However with trains travelling at ten miles an hour, there is not likely to be much danger to the public.

'But to resume our account of the trip. It was a delightful morning, and all nature looked fresh and beautiful after the refreshing rain of the previous week. We passed through the fields where the fast-growing corn gave promise of a bountiful harvest, through rich pasture lands, and through thickly wooded copses where flowers made a luxuriant carpet. In many places the line runs along the valley with woods sloping down on either side, and in the trees and hedges the birds sang merrily, and here and there a rabbit, frightened by the unusual noise of a train, ran as if for its life. Some women weeding in the cornfields paused for a few minutes and waved their weed-hooks in the air, the cows grazing in the meadows stood still and gazed in blank amazement, while the horses took to their heels. From the few straggling cottages on the side of the line women and children looked out and waved hats and handkerchiefs, and from the woodhouse of one of the cottages a flag was flying. Men and youths, hearing the train coming, left their work and posted themselves on the fence to have a look at the first train on their new line. At Bentworth-Lasham station quite a dozen women and children, in aprons and pinafores, had assembled and cheered as we steamed slowly into the station. Here we dropped two or three passengers and picked up as many more, but as there was no shunting to be done here, not till we came back, the wait was not a long one. Among the passengers picked up at this station was the Rev. W.G. Cazalet, Rector of Bentworth, who happened to get into the compartment in which the writer was travelling, and whom the latter has to thank for much information, imparted in the most pleasant manner, as to the localities through which we passed during the remainder of the journey. Away on the hills to the right was Burkham House, the residence of Mr. A.F.Jeffreys MP., a little further on was Bentworth Lodge, where resides Captain Stephens and whose lodge gates are in close proximity to the railway. Still further on, on the same side of the road, but like the other places a good distance off, is Thedden Grange whilst on the left, up among the hills is Shalden Manor, the residence of Mr. J. Gathorne Wood. Adjoining Mr. Wood's estate on the north side of the railway is a well-timbered forest sloping down in majestic grandeur to the line, and this Mr. Cazalet mentioned is the property of Winchester College, who also own a considerable tract of land on the other side of the line, from which a large quantity of gravel has been obtained by the constructors of the railway. Beyond this we pass the district of Beech, where the people live in bungalows, and then we came to the A.M.B. Hospital Camp, which at present has a very absent-minded appearance. And then by a gentle curve we came on to the Winchester-Guildford line, and running past the hop-fields, where the hops are fast making a good show, we pulled up at Alton at twelve minutes past ten, having accomplished the journey of a little over thirteen miles in three minutes less than an hour!

'After all the enthusiasm which the good people of Alton had shown with regard to the new railway, with its promise of a flourishing market, and an almost mushroom like growth of the town, one expected to see half the townspeople, together with the leading members of the District Council, at the station to greet the first train. But the modesty which characterised the LSWR Company seems to have infected the people of Alton as well, for there were not a score on the platform, and no one raised the cheer which the trippers considered their endurance and pluck deserved. As the train was timed to leave Alton on the return journey at 10.25 there was not time to explore the neighbourhood or even pay a visit to the nearest hotel, consequently the passengers waited in the vicinity of the station.

'Practically everyone who had made the whole of the initial journey returned on the train to Basingstoke, as well as those who had been picked up at intermediate stations, and in addition there were three or four Altonians who indulged in a ride to Basingstoke. The train left at 10.30 am five minutes late, but the return journey was accomplished in less time, there being no long waits. The longest this time was at Bentworth-Lasham, where there was an old coach to be picked up from the siding. At Herriard pretty nearly a dozen passengers joined the train, these being principally village women with their market baskets. At a-quarter-past eleven we arrived safely back at Basingstoke, the entire journey having been accomplished without any mishap, and having been much enjoyed by the passengers.

On the second train for Alton which left Basingstoke at 12.20, the Altonians who had come with us returned, and there were a few other paying passengers, making the total booked on both trains from Basingstoke nineteen. To this second train was added a saloon coach, in which rode several of the directors and chief officials of the line, including Colonel the Hon. H.W. Campbell (Chairman), Sir Charles Scotter (Deputy Chairman), Sir Wyndham Portal, D.L., the Rt. Hon. W.W.B. Beach, MP, Mr. Spencer Portal, Mr. C.J. Owens (General Manager) and Mr. Sam Fay (Traffic Superintendent). At Bentworth station these were joined by Mr A.F. Jeffreys MP and Miss Jeffreys, who accompanied them to Alton. From Alton the directors' party returned by a special train to Bentworth station, whence they drove to Burkham House where they were the guests of Mr. A.F. Jeffreys and Mrs. Jeffreys at luncheon. Later in the day the special took them back to Basingstoke.'

The guest list is of interest in itself. Jeffreys who had done much to support light railways seems to have overcome his frustration at the slowness of the line's opening whilst Scotter had always been the force behind the line. Sam Fay who started his railway career at Itchen Abbas on the Mid Hants line was soon to leave the LSWR. Seven months later he took the post of General Manager on the Great Central Railway and became one of the country's truly great railway managers. Retiring in 1923 at the Grouping, he returned to his native Hampshire dying there in 1953 at the age of 97.

'The South Western Gazette' also included a report of the day's events although the article contained some appalling inaccuracies, referring to standard gauge of 5ft. 8$^{1}/_{2}$in. and a station at Herriam. If the official staff magazine of the railway company could not get it right what chance was there for the local newspaper? No mention was made of the final cost of the line which totalled £107, 842 .14s 9d. As to the first day's receipts, these can only be described as small; 15 tickets were issued at Cliddesden, the 24 at Herriard yielded about 35/- whilst the receipts at Bentworth and Lasham totalled 18/10d.

Power for the first train was provided by a LSWR '02' class 0-4-4 tank engine, attached to which were the vehicles described in the newspaper article. Traffic must have improved quite quickly for two weeks after opening the Alton local newspaper reported that "the traffic is quite as good as the company expected and it will probably be necessary in the near future to add another coach to the train" and drew attention to the wool dealers at Basingstoke and Alton who would "this year find a great advantage in the new line". It was also reported that the cattle grids which had caused so many problems to the local councils had "been put to the test" with the result that "some of Squire Jervoise's sheep have broken their legs".

On 27th June, a few weeks after the opening the LSWR wrote to the Board of Trade informing them that all the requirements contained in Major Pringle's report had been complied with. This included the removal of the points to the hospital at Alton which went out of use once the Boer War ended. Re-inspection was carried out on 30th September without any problems. Nevertheless Basingstoke UDC were still unhappy, for on 18th December they complained about the condition in which Worting Road was left by the contractor.

With the opening of the new line the LSWR issued the necessary instructions as to its working. These included the head codes to be used;

"By day - Diamond on the near side of smokebox and diamond at the centre of the buffer beam.

By night - Green light as per above positions.

Formation of train - The accommodation for passengers will consist of three four wheeled vehicles, viz. one third, one comp. and one van. The load of a train must not exceed ten loaded wagons or fifteen empty wagons in addition to the before mentioned coaches.

Again reputed to be taken on opening day 1st June 1901, this was the photograph that was reproduced as a woodcut by the Hants & Berks Gazette on 8th June. The view clearly shows the sparse nature of the countryside and the singular lack of passengers. To the left is the wind pump and water tank behind which were the cottages. To the right behind the trellis fence was the small goods yard. The formation of the train, clearly mixed, corresponds with the description in the contemporary newspaper article.

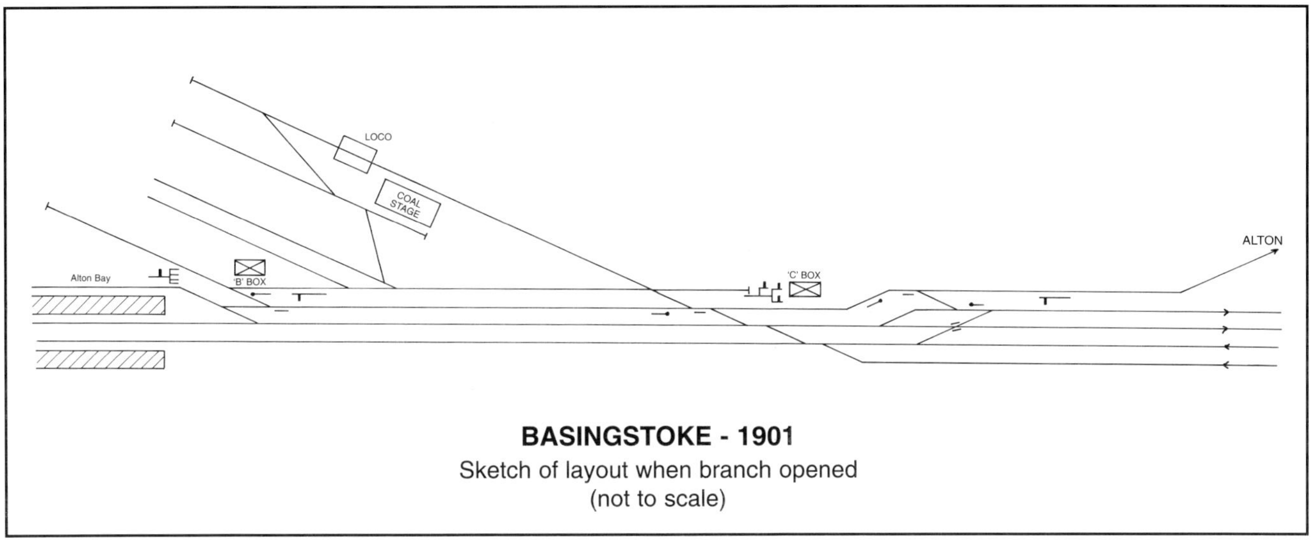

The track layout at Basingstoke at the time of the opening of the line but before the station was remodelled.

Locomotive power for the train services came from the shed at Basingstoke. Mostly these were Adams' 0-4-4 tank engines of the '02' class, number 203 soon becoming the regular engine. Occasionally a '395' class 'Jumbo' 0-6-0 was used. These were only permitted to travel smokebox first on account of the numerous occupation and level crossings. Turntables existed at both terminal stations thus ensuring the rule was carried out. As a rule tank engines left Basingstoke smokebox first and were not turned at Alton, returning with the bunker leading.

The initial train service as reported in the newspaper, was three trains each way on weekdays. The Alton newspaper's report was either accurate or prophetic for from October 1901 the service increased to four trains. All of these ran as mixed services if required. Even so the LSWR were able to work the line using a single locomotive. This may have led to economical methods of operation but it did little to provide a service that would stimulate the development of the area. With no services scheduled to cross at Herriard it is doubtful whether the passing loop was used to any great extent at the period. It was inevitable that accidents would happen, the first mishap occurring on 5th October 1901 when Cliddesden level crossing gates were demolished. This

The west end of the re-constructed Basingstoke station with 'T3' 4-4-0 No. 574 standing in the down slow platform with a Southampton service. To the right is the Alton bay with the station buildings beyond. (*Lens of Sutton*).

Exterior of the reconstructed down side at Basingstoke station which was clearly much more spacious than its predecessor. The difference in levels is apparent with passengers having to surmount a flight of steps upon leaving the booking hall to reach the platforms. The view was taken in 1925 and although the station is still in LSWR colours the posters are in the new corporate style of the Southern Railway.

A view of the Alton bay after Basingstoke station had been reconstructed. The van is hard up against the buffer stops.

(Lens of Sutton)

was followed by an accident on 6th March 1903 when a locomotive derailed whilst working a ballast train as it approached Butts Junction. As a result the line was blocked for 4½ hours.

To encourage local traffic cheap day return tickets were introduced at an early date. These were as follows:-

Basingstoke -	Cliddesden	6d
"	Herriard	9d
"	Bentworth & Lasham	1/-
"	Alton	1/6
Alton	Bentworth & Lasham	7d
"	Herriard	10d
"	Cliddesden	1/1
"	Basingstoke	1/6

Today these fares seem ridiculous so they must be seen in their historical context. In 1901 H. Rider Haggard visited the area meeting the owner of the Herriard Estate, F.M.E. Jervoise. He came to the conclusion that day labourers earned 13/- a week, but "other extras bring wages up to 16/- a week when in regular employ". Carters' wages were put at 19/- to 20/- a week. Thus, in spite of these cheaper fares, transport by train would have been an expensive luxury for the majority of the local population.

No doubt the operating costs compared with the limited receipts obtained caused the LSWR to investigate cost cutting initiatives. On 4th October 1903 a recommendation was made to the Board that the line should be worked by steam railmotors, the cost of the units being £1,380. Railmotors were an Edwardian innovation introduced initially to combat the inroads into traffic receipts caused by the development of the electric tram. The LSWR had pioneered their use, with the first pair being introduced on the Fratton to East Southsea line in conjunction with the LBSCR in early June 1903. The company recognised that the units would serve lightly used branch lines equally well, hence the proposal to use them on the Light Railway. Although the idea was initially deferred, Board approval was given for the construction of two units in December 1903. The railmotors were built at the LSWR's Nine Elms works and entered service on 1st July 1904. Given the title of Class 'H12' the layout comprised a driving cab, a first class saloon for eight passengers and a third class saloon for thirty two. The entrance was between the two saloons whilst at the far end was the cramped luggage, guard and alternative driving compartments. It would appear that they worked the same schedules as previously used by locomotive hauled trains, but it soon became apparent that the sharp curves and steep gradients of the line coupled with the problem of working the goods traffic were too much for them. Frequently the motors would stall near Cliddesden - the service then being rescued from Basingstoke. Although fuel costs were lower than conventional trains (in 1906 14 lb of coal per train/mile as opposed to 47 lb) this could not save them. Accordingly on 12th August 1904 both were transferred away to work the Bishops Waltham and Turnchapel branches which were less steeply graded.

BASINGSTOKE AND ALTON LIGHT RAILWAY.

TRAIN SERVICE—WEEK DAYS. NO SERVICE ON SUNDAYS. T T 2/69,016.

Distance.			1 Mixed Train.		2 Mixed Train.		3 Mixed Train.		4 Mixed Train.		5			
M.	C.		arr.	dep.	arr.	dep.	arr.	dep.	arr.	dep.				
			a.m.	a.m.	a.m.	a.m.	p.m.	p.m.	p.m.	p.m.				
—	—	Waterloo	...	5 50	...	9 0	...	12 50	...	4 55	...	...	...	...
—	—	Basingstoke	...	7 35	...	10 35	...	2 38	...	6 5	...	...	...	...
3	2	Cliddesden...	7 46	7 47	10 46	10 47	2 49	2 50	6 16	6 17	...	...	...	...
6	46	Herriard	7 55	7 56	10 55	10 56	2 58	2 59	6 25	6 26	...	...	...	...
9	19	Bentworth and Lasham	8 4	8 5	11 4	11 5	3 7	3 8	6 34	6 35	...	...	...	...
13	4	Butts Junction	8 16		11 16		3 19		6 46		...	...	...	...
14	14	Alton	8 19	...	11 19	...	3 22	...	6 49	...	...	...	...	...
—	—	Waterloo	10 16	...	1 0	...	5 30	...	8 46	...	...	...	...	...

Distance.			1 Mixed Train.		2 Mixed Train.		3 Mixed Train.		4 Mixed Train.		5			
M.	C.		arr.	dep.	arr.	dep.	arr.	dep.	arr.	dep.				
			a.m.	a.m.	p.m.	a.m.	p.m.	p.m.	p.m.	p.m.				
—	—	Waterloo	...	6 0	...	10 5	...	2 45	...	5 5	...	...	...	...
—	—	Alton	...	8 50	...	12 5	...	4 55	...	7 5	...	...	...	...
1	10	Butts Junction	8 53		12 8		4 58		7 8		...	...	...	...
4	75	Bentworth and Lasham	9 5	9 6	12 20	12 21	5 10	5 15	7 20	7 24	...	...	...	...
7	48	Herriard	9 14	9 15	12 29	12 30	5 23	5 28	7 32	7 36	...	...	...	...
			Cross No. 2 Down.											
11	12	Cliddesden	9 23	9 24	12 38	12 39	5 36	5 41	7 44	7 48	...	...	...	...
14	14	Basingstoke	9 35	...	12 50	...	5 53	...	7 59	...	...	...	...	...
—	—	Waterloo	11 5	...	2 52	...	7 30	...	10 6	...	...	...	...	...

All Trains must slacken speed at Butts Junction to admit of the Exchange of Train Tablet.

The timetable for the line - 1st October 1903 to 31st May 1904.

On 5th October 1904 the LSWR Locomotive Committee were informed that they had been withdrawn from service on the branch "owing to insufficient accommodation and lack of ventilation" and that ordinary working had resumed. However the LSWR built a further thirteen railmotors in 1905 and 1906. Classified 'H13' it has been suggested that all of the motors worked 'their trial one thousand miles on the light railway, in some cases coupled together.' Unfortunately no photographic evidence exists. Overall the experiment with railmotors was not a success. Their unsuitability was finally reported to the Traffic Committee in October 1916 when it was minuted that they could not be economically used on any section of the railway except between Wadebridge and Bodmin. Consequently 12 out of the 15 were converted to open corridor coaches.

Following the departure of the railmotors, services were increased to five passenger trains each way from 1st October 1904 together with a separate goods working. In addition there was a cattle train which ran on Wednesdays as required, intended to serve Basingstoke livestock market. From June 1905 further improvements took the number to its peak of six passenger trains each way; a level at which it remained until January 1916. These improvements had only been achieved at the expense of using an additional crew to work either the early or late trains plus the separate goods workings. Consequently operating costs increased accordingly.

The re-instatement of locomotive hauled workings led the LSWR to request the Light Railway Commissioners in November 1904 to raise the line speed from 20 to 30 mph, the reason given being the desire to improve the passenger service. The railway was partially successful and the line speed was increased to 25 mph. Objections were received from Basingstoke UDC who stated that they wanted gates at Viables crossing, but their request was refused. The alteration to the line speed was covered by the Basingstoke and Alton Light Railway (Speed Amendment) Order of 26th January 1905 which was confirmed by the Board of Trade on 6th April. The result was not only the improvement in services noted above but a reduction in the time taken by passenger trains. On opening in 1901, 53 minutes had been allowed for the complete journey; by June 1905 this was reduced to 44 minutes.

From an unsteady start, the traffic receipts slowly began to increase. Outgoing traffic reflected the area's agricultural base with livestock, timber, timber products and milk – the latter being the most important. Inward traffic was varied but with coal and building materials being the main items. Herriard Estate generated much business with a wide range of products being carried. Oil, manure, paint, seeds, fencing, bulbs, hardware, cement, tiles, glass and even a dog from Overton were items carried in the first year and show the railway acting as a general carrier. Even so the branch was not a paying proposition, but it seems that the LSWR were prepared to support the line with an adequate service subsidised by its more profitable operations elsewhere.

The years from 1905 onwards are conspicuous by a lack of reference to the line in official sources. Nothing untoward seems to have happened until 29th February 1908 when the 10.39 am from Alton ran into a pair of horses drawing a wagon across an ungated crossing between Herriard and Cliddesden. In such a one sided contest, it is not surprising to record there was no damage to the train. Unfortunately one horse was killed, the other injured and the wagon damaged.

On a line beset by many crossings, it was inevitable that a further incident of this nature would occur. On 5th August 1911 the 9.35 am from Basingstoke violently collided with a farm cart at Viables Crossing, Basingstoke being driven by Mr. S.C. Gear the farmer at Viables Farm. The injuries sustained by Mr. Gear were serious and proved fatal. The incident was difficult to explain as he was familiar with the crossing but nevertheless it did re-enforce the point made by Basingstoke UDC six years before when they asked for the crossing to be gated.

The final incident occurred at Cliddesden station on 23rd November 1912 when a passenger, Richard Cole attempted to re-enter the train he had just vacated. He was a 46 year old farmer from Home Farm, Farleigh Wallop who had made a day visit from

A later view of Thornycroft's works taken c.1919 with the railway in the centre of the picture behind the trees, probably those planted by the LSWR in 1906.

A heavily retouched but nevertheless interesting early view of Thornycroft's works facing the Worting road. The siding connection can be seen curving into the light railway which runs left to right across open country.

Cliddesden to Ash. He suddenly remembered he was being met at Basingstoke and tried to rectify his mistake with tragic consequences. Unfortunately the train was on the move and he clung to the side of the carriage as he tried to open a door. He struck the sturdy level crossing post at the platform end and was crushed, breaking his neck and fracturing his skull before he fell beneath the train. He left a wife and six children. This proved to be the only passenger fatality the line ever had. However these somewhat gruesome accidents must be seen as isolated incidents for they fail to convey the timeless quality of the branch at the zenith of railway operations in the nation as a whole.

There were two developments to the track layout prior to 1914. In March 1904 a new siding loop was installed at Bentworth and Lasham - details are given in chapter 10 on the station.

The other alteration occurred near Alton where the hospital siding, which had remained out of use since the line's opening, was reconnected. The hospital huts had stood empty for many years until they were brought back into use as a children's hospital, primarily for those suffering from tuberculosis. Renamed the Lord Mayor Treloar's Hospital (after the Lord Mayor of London) the first children arrived on 7th September 1908. It was the need to provide direct access for coal wagons to the hospital power plant that led to the request for the siding to be reconnected. The cost was estimated at £820 of which £529 was for work within the existing railway boundary - largely track work and signalling alterations. Worked by a reversible steam winch attached to the power house, it was stated that this would "minimise manual labour and relieve the horse." At the same time it was appreciated that improved passenger access was needed for both patients and visitors. Consequently a request was made for a private platform with access to the hospital only. This was agreed to by the LSWR, the cost of provision, £170, being paid for by the hospital authorities.

Inspection of the new facilities took place by Major Pringle of the Board of Trade on 14th April 1910. His report said:-

> "A new siding has been laid on the south side of the single line between Alton and Basingstoke at Sir William Treloar's Home for Cripples. The points face trains from Alton and are equipped with facing pointlock and detector. They are worked from a new ground frame with two levers in use which is controlled by the electric tablet for the section (Butts Junction-Herriard). The interlocking is correct.
>
> The up home signal (which is close to the new connection) for Butts Junction is worked from the junction signal box, where there is no alteration. This signal cannot be lowered unless the points to the new siding are in their proper position owing to the means of detection provided. It would have been more correct to have provided a lever in the ground frame to slot this home signal or to bolt the points from the junction signal box, but I think the arrangement adopted may be accepted as providing sufficient security.
>
> A small platform for the use only of the patients at the Home has been erected...."

Railcar No.1 seen at Cliddesden in the summer of 1904. Judging by the condition of the driver this was probably a test run with a fitter driving. Behind the vehicle are the staff cottages and the name of the station is neatly picked out in chalk. It was at the level crossing shown here that the only passenger fatality occurred.

Companion vehicle No. 2 working a service train between 1st July and 12th August 1904. To the right is Mr. Bushnell. These two photographs show the two different driving ends and it would be interesting to know whether there was any appreciable difference if the locomotive pushed or pulled.

Richard Cole and his wife Laura with five of their six children. (left to right), Alison, Richard, Hilda, Lily and Nellie. c.1906. *(Courtesy Richard Hooper).*

The platform, called Alton Park, did not appear in public timetables. Passengers were catered for every Thursday when the 1.36 p.m. from Alton made an additional stop at the platform. Return traffic was dealt with by the 4.20 p.m. from Basingstoke. By the end of 1911 cheap tickets were introduced for this traffic, a facility that was withdrawn, never to be resumed in July 1915 "owing to the exigencies of military requirements". Meanwhile the siding handled up to 1500 tons of coal annually. Initially it was worked by the weekday pick-up goods, although in later years trip working from Alton was resorted to.

The last years of the Edwardian era are generally regarded as the golden age of railway travel and apart from the accidents previously recorded, the branch had an uneventful existence. It was that low level of activity that led to the line losing more than £4,000 in 1913. The coming of war in 1914 had little immediate effect apart from the fact that some railwaymen from Basingstoke and Alton were quick to enlist. Indeed it was not until January 1916 that services were reduced with passenger trains reverting to four trains in each direction with an additional mixed train on Wednesdays and Saturdays. Freight services were unchanged.

On the outbreak of The Great War all of Britain's railways were put under Government control. Day to day management was entrusted to the Railway Executive Committee (REC), a body composed of the General Managers of the major railway companies. Although the Chairman of the Committee was the President of the Board of Trade, operations in reality were controlled by the Acting Chairman, Herbert Walker, General Manager of the LSWR.

Herbert Ashcombe Walker had joined the LSWR as General Manager on 1st January 1912 and was without a doubt an outstanding railway manager. It was this man, more than any other, who was to control the fate of the Basingstoke and Alton Light Railway. If Scotter had been largely instrumental in getting the line built, it was Walker who was to make the decisions that had a lasting effect on the line's history.

By 1916 the war was putting a severe strain on the resources of the railway industry and restrictions on passenger services were imposed to conserve manpower and equipment. However it was a shortage of permanent way materials, especially rails for use in France that led the REC to write to the railway companies on 4th October. They were asked for second hand materials or whether they "had any sections of line which could be thrown out of use so that the rails might

Patients de-training at the private hospital platform known as Alton Park which served Treloar's Hospital. From here a path ran to the main hospital complex. The private siding was out of sight beyond the platform. *(Lens of Sutton).*

A later view of Alton Park with the platform now faced in concrete and the whole resurfaced. The platform nameboard has changed and the temporary wooden platform has been replaced. The check rail was necessary because of the sharp 10 chain radius curve at this point whilst the signal is Butts Junction No. 37, situated 185 yards from the actual junction.

Part of the Lord Mayor Treloar's Hospital soon after opening. The Basingstoke and Alton line is seen curving around to the left of the complex whilst the Mid Hants line runs straight across the middle of the picture from right to left. A puff of smoke denotes a train approaching Butts Junction from the Alton direction.

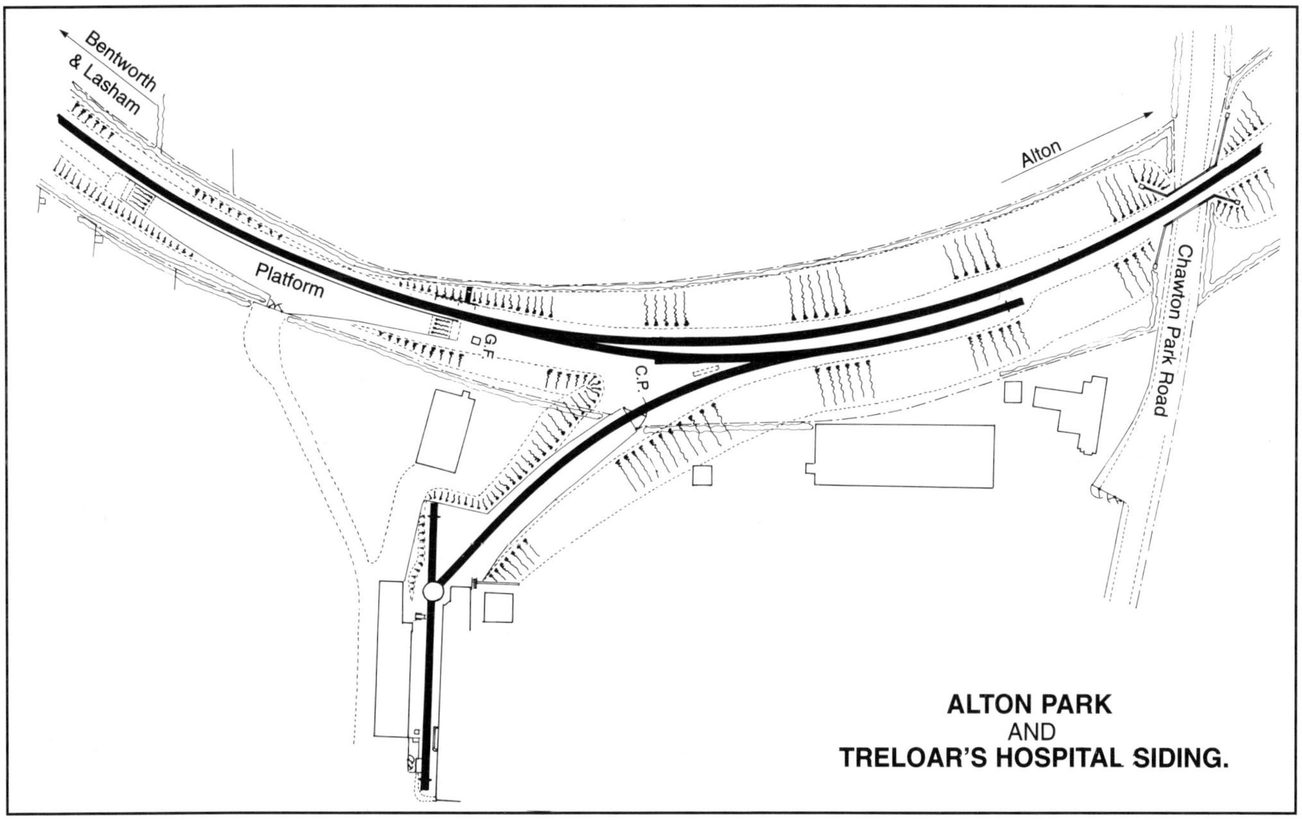

be recovered and sold to the War Department." Unable to get the desired response additional pressure was applied in early December with the demand for fare increases, reduced passenger services and the closure of branch lines. Two hundred miles of materials were wanted and each railway company was given a figure based on the size of their system.

The LSWR share was set at twelve miles and realising that his company must lead the way in this matter, Walker decided the required material could be found by closing the Basingstoke and Alton branch. The obvious question arises as to the motives behind the decision. Certainly it was a convenient excuse to curtail services on what was an unprofitable line, but also the action was probably forced on Walker because of his position within the REC . Whether the twelve miles demanded by the REC . were made by Walker in the first place so that the Basingstoke and Alton would fit them is a moot point that will always remain unanswered.

In spite of these questions the result was the same, with the local population being given *only one week's notice* of the termination of the railway service. Officially the line was to be "temporarily closed", with the last trains running on Saturday 30th December 1916 "until further notice". The railway had been in operation barely fifteen years. That everything had been carried out with haste cannot be doubted. Although the LSWR Engineering Committee were informed on 21st December 1916, the Board as a whole were not formally informed until 23rd January 1917. Meanwhile Walker wrote to the President of the Board of Trade seeking his authority for this course of action and to indemnify the railway from action that may result. This was duly given.

Public reaction to the closure was philosophical, the Basingstoke local paper accepting that the step had been "taken solely in the national interest" even though "great inconvenience to the residents of the villages served" would result. Interests in Alton, aware of the growth and importance of Basingstoke in the last few decades looked upon it as an "opportunity for the town to make itself a centre for the collection and distribution of produce."

If some doubt exists about the reasons for the line's construction, there can be few surrounding its closure. At one stroke Walker had eradicated a loss-making service whilst realising the full market value of the assets employed. He even negotiated for the Government to pay for the lifting of the track - a cost his company would normally have to bear. The whole was an excellent business proposition. At a time when the LSWR had other loss generating lines (Hurstbourne-Fullerton and Ringwood-Christchurch being two obvious examples), the reason for the selection of the Basingstoke and Alton was probably its proximity to the port of Southampton and the fact that its twelve mile length just fitted the R.E.C. demand. Certainly Walker had to show that his company was prepared to take the initiative on certain issues. The closure satisfied everyone, apart from the small communities it served.

Maintaining the railway in its early years are (left to right) Messrs. Oliver, Archer, Bradley and Forder. The Oliver family were employed extensively on the line and lived in the cottages at both Herriard and Bentworth & Lasham. It was Forder's wife who was injured whilst travelling on the rail replacement bus service on 1 December 1917 subsequently receiving £5 in compensation.

Nevertheless there was a problem. The nation was desperately trying to maximise agricultural output yet the LSWR had just closed a line (albeit with Government approval) which served a totally agricultural district. To overcome this difficulty a motor lorry service was instituted from Herriard and Cliddesden to Basingstoke which would cater for "goods, parcels, milk and other perishable or important traffic". Herriard station was to open from 7.30 a.m. to 7.30 p.m. on weekdays and Cliddesden from 9 a.m. to 6 p.m. No mention was made of Bentworth and Lasham although the South Western Railway Magazine for February 1917 still listed it as being staffed. By January 1918 the same magazine reported that both Cliddesden and Bentworth and Lasham were closed. Passengers were not catered for being left to fend for themselves as best they could. It would appear that railway staff and their families were exempt from this ruling, for on 1st December 1917 the wife of a platelayer was injured (and later compensated) "whilst travelling on one of the company's motors from Basingstoke to Herriard".

The vehicle used for the service was a 'Clayton' chain driven solid tyred omnibus, registration number CX 1682. Built in January 1915 it cost £734. 15/- and was one of two ordered to operate the Exeter to Chagford service. The bus body was now removed and replaced by a new commercial cab and body. The vehicle was kept in a "temporarily built" garage on a sleeper base at Herriard station which was provided at a cost of £130. Meanwhile the LSWR lifted the track in 1917, being paid £26,363.7.5d of which £2,452.14.6d was in respect of the lifting and handling of the materials. At the time they also applied for a rates reduction now that the service was "suspended" and were successful, previous costs of £757 per annum being reduced to £219.15/-.

The route and the railway could now be described as being in a state of limbo. The line's commercial operations were left in the hands of the Herriard stationmaster who dealt with "the special needs of the district". However the sidings to Treloars at Alton and to Thornycroft's at Basingstoke were retained as both establishments were necessary to the war effort. The line continued beyond the latter's siding to serve a War Department hay dump. Protected by a rail scotch 500 feet from the stop blocks it could contain a maximum of 25 wagons. To serve the dump, trains were propelled from Basingstoke whilst the shunter in charge had to secure the brakes on the leading wagons to control the train on the initial falling gradient. Once inside the rail scotch (which was fitted with a padlock) all the wagons brakes had to be pinned down as the line was on a 1 in 50 gradient.

When the wagons were hauled to Basingstoke the shunter was instructed to ride on the rearmost vehicle, presumably an open wagon. Services on the branch were now basic to say the least!

The downside yard and timber stacking ground at Basingstoke West yard in 1925. To the right of the corrugated hut are the running lines as follows:- light railway, 'down' slow, 'down' fast, 'up' fast and 'up' slow.

A

PUBLIC MEETING

WILL BE HELD IN THE

TOWN HALL,

BASINGSTOKE,

ON

Wednesday, 10th Jan., 1923,

TO PROTEST

Against the proposed abandonment of the

BASINGSTOKE & ALTON LIGHT RAILWAY,

And to decide what action shall be taken.

The Chair will be taken at 3 p.m., by

HIS WORSHIP THE MAYOR OF BASINGSTOKE,

(MR. ALDERMAN J. W. MUSSELLWHITE).

Supported by The COUNTESS OF PORTSMOUTH, Col. Sir ARTHUR HOLBROOK, K.B.E., M.P., Col. W. G. NICHOLSON, M.P., and other Ladies and Gentlemen.

A. KNIGHT, PRINTER, BASINGSTOKE.

Poster advertising the public meeting on 10th January 1923.

CHAPTER 6
THE FIGHT FOR REINSTATEMENT

During the period of hostilities there were no complaints from the local population over the curtailment of their transport facilities who no doubt saw the move as a necessary part of the national war effort. Nevertheless it was confidently expected that services would be restored with a return to peace, after all the line was only "temporarily closed." Indeed, hardly had the war ended, when on 23rd January 1919 the directors of the LSWR received a petition from "residents in villages on or near the line" asking for the "restoration of passenger and goods services with the least possible delay". The directors politely refused this request and on 25th March 1920 Walker informed the Ministry of Transport that it was unable to see its way to "re-establish the service". This decision could be made with impunity for Walker was only too well aware of the indemnity the railway companies had received from the Board of Trade in 1916. Likewise the LSWR also refused to re-introduce cheap tickets to Treloar's Hospital.

Needless to say local concern continued, so much so that the then MP for Basingstoke, Sir Arthur Holbrook, asked the House of Commons on 25th November 1920 when the service would be restored. In reply the Minister of Transport said that he "would not be justified in giving a direction to open the line" due to the cost of restoration "and a heavy annual loss in working which during the period of Government control would have to be met by government guarantee" (at this time the railway network was still under Government control). Clearly the Minister was not interested in what was seen as a parochial matter which could cost the Exchequer a considerable sum. The final part of the Minister's answer that the Ministry was looking as to whether "the needs of the district may be supplied in another manner" was just a way of pacifying Holbrook.

Whilst this political agitation was taking place the LSWR started to put its own house into order. At the main Board meeting on 13th October 1921 Walker set out the financial facts relating to the Light Railway. The line had cost a total of £107,842.14.9d to build against which the company had recovered £23,910.12.11d from the government for the permanent way materials sold to them in 1917. The balance on capital account was £83,932.1.10 which was represented by the two short sections of track still in use, twelve miles of trackbed and seventeen staff houses. In 1913 receipts were put at £1,232 whilst operational costs had been £5,400. Not surprisingly Walker's recommendation that the line should not be re-opened was accepted. The decision was not made public and instead matters were held in abeyance pending the preparation of the necessary Parliamentary Bill.

During this period the LSWR were planning other cost reduction exercises. With the abandonment of the Basingstoke branch it was proposed to abolish Butts Junction signal box thereby saving £470 per annum. The Meon Valley and Mid Hants lines were to be worked as single independent lines from Alton with the provision of a scissors crossover at the station whilst the tablet instruments would be re-located to the station signal box. However the changes, which were costed at £5,000, could not be put into effect until the closure of the Light Railway had been confirmed by Parliament. Attention also focused on the replacement road motor service. In May 1922 it was reported as being unremunerative and the LSWR Traffic Committee recommended that the service be terminated. Mr. E.E. Gilbert of George Street Basingstoke was to be appointed as the railway company's delivery agent and at Herriard the Station Master's house, station buildings, yard and garage were let to him for an annual rent of £70. As a result of this move savings of £843 p.a. were expected with the changes being instituted from 1st January 1923. Thus, having solved the problem of providing a freight service to the area at no cost to itself, the LSWR then sought the legal sanction to close the line. Unfortunately the LSWR had no Parliamentary Bills scheduled for 1922, so the matter was conveniently handed over to the newly formed Southern Railway into which the LSWR had been amalgamated from 1st January 1923.

With no effort being made to restore train services, rumours began to circulate that the closure of the line was permanent. These reached a peak during the latter part of 1922. Again the matter was taken up by Sir Arthur Holbrook, who having failed to receive any satisfactory assurances from the railway company, attempted to provoke discussion again by asking the Parliamentary Secretary to the Minister of Transport whether the Government would consider restoring the line as an unemployment measure. Not surprisingly the request was refused.

However, it was the presentation in late 1922 of the Parliamentary Bill to close the railway that galvanised the ineffective protest movement that had existed since 1919 into a highly motivated pressure group. Led by Major F.H.T. Jervoise of Herriard Park the support of the local gentry, landowners, politicians and businessmen was soon enlisted. In early December 1922 Jervoise addressed the Hampshire National Farmers Union whilst later in the month a circular letter in his name called the attention of the public to a meeting that the Mayor of Basingstoke was holding on 10th January 1923. Meanwhile on 1st January 1923 Jervoise addressed the annual meeting of the Hampshire branch of the Central Landowners' Association. His condemnation of the railway company was strongly supported by Major General

Jeffreys (son of the late A.F. Jeffreys MP) and the Countess of Portsmouth. Criticisms of the railway company were varied - they did not encourage through traffic, there were not enough halts, the area had come to depend on the line and now the "countryside had an ugly scar drawn across it". Somewhat unrealistically the opinion was expressed that perhaps the GWR would work the line! The public still had perceptions of old rivalries that no longer existed.

The public meeting on 10th January 1923 was well filled with local landowners, representatives of the local authorities, Basingstoke Chamber of Commerce and the National Farmers Union. In all seventy three people attended, but what they lacked in numbers was certainly offset by the influence they had. As can be expected there was a lively discussion which somewhat exaggerated the local conditions; it was an omen of what was to follow. Overall the meeting expressed the opinion that "they were up against an unimaginative, unenterprising and soulless body of directors." Proposals for future action included the idea of re-instating the line as a narrow gauge railway. Henry Greenly consulting engineer to the Ravenglass and Eskdale Railway was in the audience and it is interesting to speculate on the reasons why he was there. A few years later he became directly involved in the building of the Romney, Hythe and Dymchurch Railway and there is a possibility he was looking at potential sites for the railway he was later to build in Kent.

It is hardly surprising that there was no dissent to the proposition that a Committee be formed to fight the closure proposal. Headed by Sir Arthur Holbrook MP it included landowners, Council representatives and a trade unionist as well as someone who had been connected with local railways since 1888, A.F.M. Downie from Alton. Finally a petition opposing closure was signed by everyone there. It is interesting to note that few private individuals were involved; this should not be seen as complacency but more a reflection upon the social class structure at the time. Nevertheless Major Jervoise could be satisfied with progress to date. In a little under a month he had welded together the most important forces in north Hampshire; in reality not that difficult as they were all connected through the same political party!

The local authorities, which had proved so obstructive when the line was being built, pooled resources to save the railway they had so often criticised. Accordingly the necessary solicitors and parliamentary agents were instructed to act on their behalf. The first meeting between the two took place on 19th January 1923, and so began the official opposition to the Bill. In all four formal petitions were presented to Parliament; one came from Major Jervoise, the others coming from Hampshire County Council, Basingstoke UDC and a joint one from the Rural Districts of Alton and Basingstoke. All contained statements in which they alleged the decision to construct the line in the first place had been influenced by the numerous proposals to build railways in the area in the late 19th century. The importance of the railway to agriculture was stressed as was the increased transport costs that local businesses had faced. The fact that the line had never paid was felt to be an irrelevance. The evidence of local traders was attached to add support to the argument. A London hay and corn merchant claimed that increased transport costs had lost farmers in the area (and Mr. Jervoise in particular) 5/- a ton. A local tailor said the closure of the line had "seriously affected his business" and how "on several occasions he (had) found difficulty in obtaining a seat in the train when travelling between Alton and Basingstoke". Similarly a Basingstoke corn merchant claimed that the railway "was used constantly for the cartage of goods and how the opening of the line would be of immense benefit to farmers." On a different note Basingstoke UDC complained of the increased traffic on the roads as a result of the closing of the railway. In retrospect it can be seen that the claims made were somewhat exaggerated to support the argument. It is also noteworthy that the four who gave evidence did not show enough interest to sign the petition at the public meeting in January. Just as in the debates surrounding the building of the line, the arguments for re-instatement often found it convenient to blur fact and fiction.

In spite of objections in the House of Commons, the Bill authorising closure was approved. It then moved on to the House of Lords where a Select Committee was appointed to decide the issue. Chaired by Lord Kintore, it took place between 2nd and 4th May 1923. Sir Herbert Walker (as he had now become) represented the Southern Railway and correctly stated that no industry or large scale development had occurred as a result of the line. He went on to say that the line had been worked at a loss; this having grown from £1,878 per year in 1908 to £4,296 a year in 1913. In addition he put the cost of restoring the railway at £35,000, the maximum traffic receipts at £5,064 per annum, whilst operating costs were estimated at £11,139. He also made the point that much traffic had already been lost to road competition. Attempts were made to discredit his evidence, but failed; Walker was far too experienced and he had been well briefed. In particular he firmly refuted the allegation that the line was built only to keep out other railways.

The following day Major Jervoise was cross-examined. He estimated that the 1,800 acres he personally farmed despatched 30 to 40 tons of merchandise per month. To counter Walker's point about lack of development he claimed that he had been planning development at Lasham in 1908 but had "discovered difficulties" with the water supply. He "gave up the idea of development when the railway closed". He complained that the presence of cuttings and embankments together with "three to four foot of ballast" made the land "practically worthless" and "impossible to bring into cultivation". In particular Major Jervoise felt that the area was failing economically and he made the point that "what we want is our railway, not compensation". He was followed by a representative of the NFU and a succession of farmers who all spoke of the importance of the railway and how much they depended on it. This repetition became too much for the Chairman who, in exasperation exclaimed "a man comes here and says he has not got a railway and would like one; it is a pure waste of time".

From this point the debate changed significantly with the argument concentrating on why the line was built and the promises the LSWR and Scotter in particular, made in 1895 to various interested parties. This centred on the logic that the Southern Railway could hardly argue the reason for non re-instatement on financial grounds when its predecessor had stated quite publicly that it was prepared to build the line knowing full well it would not pay.

On the second day of the hearing two witnesses were produced whose testimony was to be significant. The first was Downie, the Alton solicitor who had helped promote two of the previous schemes linking Basingstoke to Alton by rail. He stated that the LSWR "strenuously resisted" all previous attempts to build the line and that Scotter's pledge was initially to build a light railway down the Meon Valley but this was subsequently changed to "a main line railway there and a light railway from Alton to Basingstoke." This point of view was supported by William Nicholson MP who was the Chairman of the company behind the 1895 scheme. He went further when he said that he had three interviews with the LSWR management in 1895 when Scotter was alleged to have stated "as regards the question of the line from Basingstoke to Alton, we realise the danger from which we have just escaped; we realise that if any other line (was) proposed to come through that part of the country the House of Commons or House of Lords would probably grant them the line. In order to prevent that danger in future we are prepared to take advantage of the Light Railways Act and we will make a light railway there". Nicholson added that part of the bargain was that the 1895 proposal would subsequently be dropped. When questioned about the financial stability of the 1895 company, Nicholson replied "financially we were very strong - we had a large amount of money at our disposal". When asked about the position taken by the GWR he stated that "they took a negative view; they would not assist us but were prepared to let us run into their station at Basingstoke". To round off his argument which contained parts of doubtful accuracy, Nicholson

Major F.H.T. Jervoise.

Herriard Park,
Basingstoke,
22nd December, 1922.

Dear Sir,

BASINGSTOKE & ALTON LIGHT RAILWAY.

As you are probably aware the London & South Western Railway Company are bringing an Omnibus Bill before the House of Commons for powers to, among other items, abandon the Basingstoke and Alton Light Railway.

The Mayor of Basingstoke is arranging a Meeting of Protest to be held at the Town Hall, Basingstoke, at 3 p.m., on Wednesday, 10th January, 1923 (should the date be altered I will let you know), at which our Member, Sir Arthur Holbrook, will be present.

May I ask you to make a special effort to be present at this Meeting to support the protest, as it is of vital importance that the abandonment of the line be opposed.

I intend, in conjunction with my neighbouring Landowners and others interested, to do all that is possible to have this line re-opened.

Yours faithfully,

F. H. T. JERVOISE.

Major Jervoise's circular letter dated 22nd December 1922.

The relaying gang in 1923. In charge was Inspector King whilst Reg Pye was a relaying ganger. Amongst other men involved with the permanent way were Peter Willis who was a ganger in both periods of operation, Harry Oliver, and E. Pink.

considered the closure of the line "a great breach of the pledge given to me by Sir Charles Scotter." Proof that all the proposals for linking Basingstoke and Portsmouth had failed on financial grounds fell on deaf ears. An unproven "honourable" agreement from a man who had died nearly thirteen years previously was overcoming every other argument.

In reply the Southern Railway informed the Committee that it had no record in writing of Scotter's alleged undertaking. Nevertheless the Southern Railway Company recognised that if the opposition's argument was true they would be directors of the railway linking Basingstoke with Alton and therefore they were prepared to hand over the physical remains of the line to that body, wipe out the capital debt as well as providing facilities at both ends of the line. Not surprisingly this somewhat tongue in cheek reposte failed to gain any support. Nicholson and Downie wanted the railway, not the loss associated with it.

At the conclusion of the hearing the consensus was that the objectors to closure had proved their case against the proposed abandonment and that if run correctly the line could pay. Lord Kintore in his summing up said;

> "The Committee are of the opinion that the clauses for abandonment must be struck out. Although we do not see our way to impose on the Southern Railway an obligation to relay and work, yet we hope that the Company will understand that, having regard to the conditions under which the light railway was originally authorised and that the responsibility rests upon railway companies to provide facilities, not necessarily at direct profit, it is the opinion of the Committee that the Railway Company should re-examine the situation in the light of the Committee's views."

As a result of these findings, the Southern Railway backed down, issuing a statement that they would restore the light railway with the reservation that the Company had the right to review the situation after ten years if they found the line was working at a loss. This was accepted.

Lest it be felt that the evidence weighs heavily against the Southern Railway, it should be recognised that in the early part of 1923 the company was not in a satisfactory position to contest such a case. The management structure was in total disarray with no less than three General Managers. Walker, in charge of only the former LSWR lines certainly felt the position he was in was untenable, and offered to resign. However he was supported by the Chairman Sir Hugh Drummond who knowing the calibre of the man put an ultimatum to the Board - if Walker went so would he. It was unfortunate to say the least that the Board meeting where everything was to be decided was on May 3rd, the second day of the Select Committee enquiry. Consequently Walker could be excused for having other things on his mind. Likewise, the Board, facing major organisational problems within the new company probably regarded the issue as relatively unimportant, hence the willingness to concede defeat.

Notwithstanding the victory, Sir Arthur Holbrook sounded a cautious note by stating that "now the residents are going to get their railway back it was up to them to use it". Unfortunately if they heard his words, they did not act upon them. The costs of fighting the Bill had not been cheap with Major Jervoise alone spending an estimated £850 in legal fees - costs which it is unlikely he ever recouped. The local authorities, mindful of their ratepayers, squabbled over the allocation of costs with the problem only being resolved as late as 1928.

True to their word the Southern Railway lost no time in honouring their commitment to relay the track and two gangs of their own men were allocated to the task. Unfortunately the number of individuals involved is not recorded. One gang were occupied in clearing the trackbed of six years' unchecked growth and repairing fences whilst the second gang relaid the sleepers and rails. At the same time the tentative plans to abolish Butts Junction signal box were cancelled. On 6th December 1923 whilst the relaying was taking place, the Southern Railway received a request from British Petroleum for a private siding facility to serve a bulk spirit depot storing spirit, kerosene and vapour. This was to be established just north of Messrs. Thornycroft's and on the opposite side of the line to the water works. This request was approved subject to BP paying £341, the estimated cost of the engineering and rail work together with £10 per annum rental in respect of easements and a further £1 per annum for land occupancy. These terms were agreed to by the respective parties and the siding was brought into use in 1924. It was however of limited size being only 50 feet in length.

SOUTHERN RAILWAY.

London (West) Divisional Operating Superintendent's Office, Waterloo Station.

Special Notice No. S.2437 L.W.D. TO ALL CONCERNED. 9th August 1924.

ON MONDAY 11TH AUGUST

(i) SPECIAL TRAIN will run as under:-

	arr. p.m.	dep. p.m.
BASINGSTOKE	A	2.30
Thornycroft's Sdg	2.34	2.37
Cliddesden	2.42	2.47
Herriard	2.57	3. 2
Bentworth & Lasham	3.10	3.14
Beech Croft	3.24	3.28
Cripples Home Plat.	3.31	3.38
Butts Jct	3.40	3.42
ALTON	3.45	C

(Beech Croft and Cripples Home Plat.: B)

"A" After arrival of 1.30pm from Waterloo.
"B" Stop 400 yards Basingstoke side of Cripples Home Sdg. to take up man with Tablet.
"C" Connect with 3.52pm from Alton.

THE TRAIN WILL BE PROPELLED FROM BASINGSTOKE TO ALTON.

BASINGSTOKE to provide train as ordered.
BASINGSTOKE LOCO - engine.

RETURN EMPTY WORKING:-

	arr. p.m.	dep. p.m.
ALTON		4†10
Butts Jct	4	13*
Bentworth & Lasham	4	25
Herriard	4	33
Cliddesden	4	43
BASINGSOKE	4.53	

PERMANENT SPEED RESTRICTIONS.
The undermentioned speed restrictions must be strictly adhered to on the Basingstoke and Alton Line.-

DOWN TRAINS Maximum speed per hour. Miles.	POINTS AT OR BETWEEN WHICH SPEED MUST BE REDUCED.	UP TRAINS. Maximum speed per hour. Miles.
10	Over curves between 57½ and 58½ mile posts, between Lasham and Butts Jct.	10
10	When approaching and within 300 yards of the following Crossings:-	10
25	Over other portions of the line.	25

Name of Crossing	Distance from London via Basingstoke Miles	Chains	Stations between
Viables	49	70	Basingstoke and Cliddesden
Bushey Warren	53	24	Cliddesden and Herriard
Grange Road	53	69	Cliddesden and Herriard
Herriard Common	55	17	Herriard and Lasham
Salter Hatch	55	77	Herriard and Lasham.

MR. HATCH, ALTON, to send a man (with the Butts Junction tablet) to meet the train at a post 400 yards the Basingstoke side of the Cripples Home Siding and take the return train to the place. (T.S.A.)

WW 4.30pm

A. H. HOYLE.
DIVISIONAL OPERATING SUPERINTENDENT.
per:-

Special train notice dated 9th August 1924 for the senior officers' inspection on 11th August.

To the outside observer there appeared to be little urgency in the relaying work and consequently Sir Arthur Holbrook MP wrote to Walker in March 1924 complaining that several of the workmen had been withdrawn and that in his opinion the work would take another two years to complete. This view was quite inaccurate and Holbrook's fears were unfounded for by May 1924 a series of internal Southern Railway memorandums were dealing with the opening, stocking and operation of the line. It is also interesting to record that when Walker replied to Holbrook he let it be known that the provision of a halt at Beech, near Alton, was receiving attention but for whatever reason, it was not proceeded with.

One of the first items to be dealt with by the railway was the arrangement of the tablet section with the traffic and engineering departments apparently in dispute over the issue. This was to drag on until July 1924 when a hand written note from an un-named individual made the obvious statement "surely the necessity for the tablet section at Herriard will depend upon the intended train service". The decision not to have a crossing place at Herriard was confirmed on 11th July 1924 in a memorandum to the Chief Engineer, Alfred Szlumper. The result was that the Basingstoke West to Butts Junction single line section was to become the longest on the Southern Railway.

Shortly afterwards on 24th July 1924, a senior officers' visit to the re-instated railway took place and although the method of travel is not recorded, it may be assumed that this was by train. Those present were:-

H.I. Bond	Divisional Engineer
C. Giles	Divisional Commercial Manager
A.H. Hoyle	Divisional Operating Superintendent
Inspector Webb	Stores Department
Inspector Case	Divisional Inspector

A comprehensive report was written although it was obvious that those involved in the inspection had not been informed of the track layout agreed with Szlumper a fortnight before! It is reproduced below:-

"We have visited the stations on the Basingstoke and Alton Light Railway, and find the work is well in progress, and there should not be any difficulty in the line being re-opened on the date suggested, viz., 18th August, provided the following details are put in order:-

1. Level Crossings Without Gates
The notices at points where the roads approach the railway require renovation, the present notices being obsolete headed 'L & SWR'

2. Lighting
The Lighting Engineer to provide the necessary platform and office lamps, and in conjunction with the engineer, a lamp box for the deposit of oil and the other accessories at each station.

3. Clocks
The provision of a clock at each station. In this connection the requirements of the Ministry of Transport in regard to the view obtainable of the clock from the platform will have to be considered.

4. Booking Offices
The provision of bars to protect the windows, and a decision where safes are to be provided or the cash sent in a locked box by the last train each day.

5. Cattle Pens
It is proposed to provide a cattle pen at each station but in view of the limited number of trucks dealt with when the line was previously in existence, it is considered that an ordinary loading bank with portable hurdles should at present suffice.

6. Cliddesden
In addition to the level crossing gates at this station the gates at Hackwood Lane Crossing have to be manipulated by the station staff. Prior to the line being closed the Station Master looked after the station gates and the Porter attended to Hackwood Lane Crossing when trains were running. If the suggestion to provide a Grade 1 Porter at each of the stations is adopted a Junior Porter or some other assistance will be required at Cliddesden to deal with Hackwood Lane Crossing.

7. Herriard
It is understood that a crossing loop will be provided at Herriard and a tablet instrument. If the line is worked by one engine only in steam the necessity for the loop, even if the tablet instrument is required in connection with Thornycroft's sidings and the Butts Junction connections, is not quite apparent.

7a The shed provided for housing the company's lorry during the time the line was closed has been leased to a Mr. Gilbert, who performs the Company's cartage of milk, but when the station is completed access can only be obtained to the shed over the sidings and it is recommended the tenancy be cancelled and the shed used for the storage of goods.

8. Station Houses
The station houses are at present occupied by persons not connected with the railway, and as there will probably be a difficulty in arranging for the Grade 1 Porters to satisfactorily perform their duties and live away from the stations the station houses should become available for their occupation.

9. Lasham and Shalden Crossings
These crossings are provided with gate houses and are in the occupation of Station Foreman Dean and Assistant Shunter Longhurst, Alton, respectively. Longhurst's wife is prepared to look after the crossing, and enquiries are being made to ascertain whether Foreman Dean's wife will undertake a similar duty.

10. Stores
The Stores Department will provide furniture and stores as shown on the accompanying list. Stationery for the three stations has already been despatched to Basingstoke, but although there is an accumulation of correspondence at each station which was left when the line was closed it will without doubt, be necessary for some new rate books classifications and other details in connection with coaching and goods traffic to be supplied. A stock of tickets should also be requisitioned together with fare lists for exhibition outside Booking Offices."

A few days later this was followed by details of the restored train service for the line;

	PASS		MIXED		PASS	
	arr.	dep	arr.	dep	arr.	dep
	a.m.	a.m.	p.m.	p.m.	p.m.	p.m.
Basingstoke	-	10.15	-	12.35	-	3.25
Cliddesden	10.24	10.25	12.44	12.49	3.34	3.35
Herriard	10.35	10.36	12.59	1.09	3.45	3.46
Bentworth & Lasham	10.44	10.45	1.17	1.22	3.54	3.55
Alton	10.59	-	1.36	-	4.09	-

	PASS		MIXED		PASS	
	arr.	dep	arr.	dep	arr.	dep
	a.m.	a.m.	p.m.	p.m.	p.m.	p.m.
Alton	-	11.15	-	2.05	-	4.25
Bentworth & Lasham	11.29	11.30	2.19	2.24	4.39	4.40
Herriard	11.38	11.39	2.32	2.42	4.48	4.49
Cliddesden	11.49	11.50	2.52	2.57	4.59	5.00
Basingstoke	11.59	-	3.06	-	5.09	-

Further loose ends were tied up on 9th August when it was agreed that telephones would be installed at Basingstoke telegraph office, Basingstoke West box, Butts Junction, Alton booking office and the three stations. On the same date a notice was issued giving details of a special train to run over the line on Monday 11th August. The purpose of this train was a further inspection visit by senior officers - certainly the Chief Engineer Alfred Szlumper was on board. What is certain is that it was not run on behalf of the Board of Trade. They had previously sanctioned the re-opening on 31st July stating that Colonel Pringle would undertake a re-inspection at a later date.

Behind the scenes, correspondence continued with the signalling and operating departments being 'locked' in discussions over possible changes to the working of Basingstoke West and Butts Junction signal boxes now that Herriard was no longer to be a crossing place. The outcome was to authorise the use of a 'warning' arrangement at both locations whereby a train might proceed with the expectation of being cautioned at the other end. By adopting this arrangement greater flexibility of working was achieved. Finally a special stores train was despatched from Basingstoke at 11 am on Friday 15th August. This called at all the stations dropping off the necessary furniture,

tickets etc. before returning empty from Alton.

Meanwhile day to day control of the line was vested in the Basingstoke Station Master, Mr. F. Gabriel who was awarded an increase in salary from £360 p.a. to £400 p.a. "in consideration of responsibility for Cliddesden, Herriard and Bentworth and Lasham". The additional responsibilities may not have been great, but it was a measure of the cost cutting exercise the Southern Railway undertook when they re-opened the line. Services were to be restored, but operating methodologies were not.

Town Hall, Basingstoke, scene of the protest meeting on 10th January 1923.

SOUTHERN RAILWAY

Signal Instruction No. 17, 1924.

INSTRUCTIONS TO ALL CONCERNED AS TO THE

RE-OPENING OF THE BASINGSTOKE AND ALTON LIGHT RAILWAY,

On Monday, 18th August, 1924.

The single line, 12 miles, 72 chains in length, formerly known as the Basingstoke and Alton line, between Basingstoke West and Butts Junction, will be re-opened for traffic on Monday, 18th August, 1924.

There are three intermediate stations on the line, viz., Cliddesden, Herriard, and Bentworth and Lasham.

The line will be worked under Tyer's Train Tablet System (No. 6 Instruments) in accordance with the regulations shewn on pages 30–47 of the South Western Section Appendix to the Book of Rules and Regulations.

The tablet section will be between Basingstoke West and Butts Junction.

The section clear but station or junction blocked "warning arrangement" will be in operation between Basingstoke West and Butts Junction for up and down trains in accordance with Regulation 11 of the Electric Train Tablet Regulations, and Rule 40, Clauses (*d*), (*e*) and (*f*) of the Book of Rules and Regulations.

Diagram of the lay-out, gradients and curves is attached hereto.

Trains from Basingstoke to Alton and *vice versa* must carry head signal No. 8.

Trains travelling from Alton to Basingstoke will be known as down trains, and those travelling from Basingstoke to Alton as up trains.

Speed restrictions.—The undermentioned speed restrictions must be strictly adhered to:—

Points at or between which speed must be reduced.			Maximum speed per hour. Down and Up trains.
Over curves between 57½ and 58½ mile posts, between Bentworth and Lasham and Butts Junction			10
When approaching and within 300 yards of the following level crossings ...			10
Name of Crossing.	Distance from London via Basingstoke.	Stations between	
Viables ...	49 m. 70 ch.	Basingstoke and Cliddesden.	
Bushey Warren	53 m. 24 ch.	Cliddesden and Herriard.	
Grange Road	53 m. 69 ch.	Cliddesden and Herriard.	
Herriard Common	55 m. 17 ch.	Herriard and Bentworth and Lasham.	
Salter Hatch	55 m. 77 ch.	Herriard and Bentworth and Lasham.	
Over other portions of the line			25

The first page from the official re-opening notice of the light railway, 18th August 1924.

SOUTHERN RAILWAY

RE-OPENING OF

BASINGSTOKE AND ALTON LINE ON WEEKDAYS.

Commencing on Monday, 18th August, and every Weekday until further notice,

The undermentioned service will operate:—

Waterloo	dep.	a.m. 8 30	a.m. 10 15	p.m. 1 30
Basingstoke	arr.	10 3	p.m. 12 7	2 28
Salisbury	dep.	8 55	a.m. 10 7	12 40
Basingstoke	arr.	9 54	11 14	1 58
Southampton West ...	dep.	8 40	10 12	2 3
Basingstoke	arr.	9 45	11 25	3 14
Basingstoke	dep.	10 15	p.m. 12 35	3 25
Cliddesden	,,	10 24	12 44	3 34
Herriard	,,	10 35	1 4	3 46
Bentworth & Lasham	,,	10 44	1 17	3 54
Alton	arr.	10 59	1 36	4 9
Alton	dep.	11 12	2 5	4 28
Waterloo	arr.	p.m. 12 40	3 36	6 16
Alton	dep.	...	3 34	5 45
Southampton West ...	arr.	...	5 10	7A5

Waterloo	dep.	a.m. 9 10	a.m. 11 20	p.m. 2 0
Alton	arr.	10 32	p.m. 1 14	3 30
Southampton West ...	dep.	9 33	12 33	...
Alton	arr.	11 6	2 2	...
Alton	dep.	11 15	2 5	4 25
Bentworth & Lasham	,,	11 29	2 19	4 39
Herriard	,,	11 38	2 32	4 48
Cliddesden	,,	11 49	2 52	4 59
Basingstoke	arr.	11 59	3 6	5 9
Basingstoke	dep.	p.m. 2 2	3 26	5 20
Waterloo	arr.	2 56	4 20	6 21
Basingstoke	dep.	12 28	4 13	5 34
Southampton West ...	arr.	1 21	5 10	6A42
Basingstoke	dep.	12 57	4 38	6 7
Salisbury	arr.	2 14	5 57	6 57

A—Southampton Terminus.

Waterloo Station, S.E. 1.
August, 1924.

H. A. WALKER, General Manager.

T. 2548/5/7824 WATERLOW & SONS LIMITED, LONDON, DUNSTABLE & WATFORD.

The public re-opening notice and accompanying timetable.

DAILY SKETCH, TUESDAY, AUGUST 19, 1924.

HERRIOT "MOBBED" IN PARIS : RUHR EVACUATION BEGUN

DAILY SKETCH

No. 4,808. Telephones { London—Holborn 6510. Manchester—City 6501. LONDON, TUESDAY, AUGUST 19, 1924. [Registered as a Newspaper. ONE PENNY.

DRESSING BY COLOUR CHART See Page 7

PEACE AT LAST: THE TRAIN IS RUNNING AT CLIDDESDEN

How Mr. Bushmell and his station—and his poultry—looked after the rails had been torn up.

When the little station at Cliddesden was reopened yesterday there was so much excitement that the Union Jack was hoisted upside down. During the war the rails between Basingstoke and Alton were taken up for use in France. The ex-stationmaster, Mr. Bushmell, who is over 71, was on the platform yesterday waving his stick when the first train came in.—(Daily Sketch exclusive.)

DOCTOR'S FATE

Dr. A. R. French, of Camberwell, who committed suicide in a fit of depression, had, the coroner was told yesterday, 3,000 panel patients, and his illness was due to overwork.

The guard yesterday congratulating the driver of the first train, who is on his old ground again.—(Daily Sketch exclusive.)

BARRISTER DIES ON HONEYMOON

Mr. Gavin William Ralston, who was married to a Russian countess refugee on August 6, collapsed on the road at Worth Matravers, near Bournemouth, where the honeymoon was being spent, and died immediately. Mr. Ralston was a Scottish barrister, aged 62. His wife's parents were killed by Bolshevists.—(Daily Sketch.)

The national press comes to Cliddesden - the re-opening of the station, August 1924.

CHAPTER 7
DECLINE WITH IMMORTALISATION

Public services re-commenced on Monday 18th August 1924 and were devoid of official ceremony. Despite this there was considerable media interest, railway re-openings being a most unusual event. At least one national newspaper covered the first day sending both a reporter and photographer together with the local papers. The 'Southern Railway Magazine' too subsequently carried a feature on the new service.

Beside the fare paying passengers, on board the first train was C. Giles the Divisional Commercial Manager for the Southern Railway. A Basingstoke crew were in charge of the train comprising driver William Hudson (who had been the fireman of the first train in 1901), fireman A. Willett and guard H. Tiller. At the head of the two coach train was 'O2' class 0-4-4 tank engine number E234. Forty passengers were estimated to be on board. An interested spectator to the day's events was 76 year old Charles Bushnell the retired former Station Master at Cliddesden, who still lived in

'02' class 0-4-4 tank engine shunting at Thornycrofts in Southern days.

'G6' class 0-6-0 tank engine No. 278 on a trip working between Thornycrofts and Basingstoke in Southern days.

the station house.

No doubt the novelty value of the railway initially encouraged travel. For example a total of 95 passengers were carried on the first train from Alton on Wednesday 20th August, the fact it was Basingstoke market day certainly helping to swell the figures. Interestingly 49 of those travelling came from Alton whilst Cliddesden could only muster two passengers. Even so, it must have been somewhat cramped in the usual two coach set! In spite of these exceptional loadings, the line still failed to pay its way. The gross receipts for all traffic in the first two weeks totalled £94.11.11d whereas wages costs alone accounted for £102.17/-. Running and maintenance costs were additional; consequently the outlook was economically gloomy from the start.

The re-instated service comprised three trains in each direction on weekdays only. The timings, given in the previous chapter, had been devised so that they could be carried out by one locomotive and set of coaches with one crew within a normal eight-hour working day. Fares were fixed as follows:-

	1st.. Single	1st Return	3rd Single	3rd Return	3rd Cheap Day Return	
Basingstoke - Alton	2/11	5/10	1/9	3/6	1/9	Tues & Sat
Cliddesden-Alton	2/4	4/8	1/5	2/10	1/5	" "
Herriard-Alton	1/7	3/2	11½d	1/11	1/-	" "
Bentworth-Alton	1/1	2/2	7½d	1/3	8d	" "
Herriard-Basingstoke	1/5	2/10	10d	1/8	10d	Wed & Sat.
Bentworth-Basingstoke	1/11	3/10	1/2	2/4	1/2	" "
Alton-Basingstoke	2/11	5/10	1/9	3/6	1/9	" "

Advertisements were placed in the local press giving fares, train times and "connections". The latter were poor. The local press were critical of the service from the beginning and this point was made to the local authorities. The comment that "one had to go back in 20 minutes or wait for hours" was quite valid. Farmers complained that the late departure of the first train from Alton was useless for the milk traffic; one complainant reporting there were 30 churns to be taken daily from Lasham alone. Another complaint concerned the fact that Herriard level crossing was closed when trains were due to arrive irrespective as to whether the train was actually running on time. Should the service be running late, the gates remained closed against road traffic for long periods. This action was probably the result of the reduced staffing levels on the line and the fact there was now no Station Master on site to directly supervise.

The matter of the poor service was taken up by the local authorities in earnest with Basingstoke RDC writing to Major General Jeffreys - son of the former Basingstoke MP who had been so supportive of the railway. Jeffreys replied enclosing a copy of his suggested timetable, the same as he had given to "the Station Master at Basingstoke and the Southern Railway's Commercial Superintendent of the Line". It must be said that the suggestions were sound, but the thinking behind them was probably not Jeffreys' - the handwriting was different! Initially the Southern Railway refused to alter the timetable as it wanted to minimise operating costs by operating within the eight hour day now worked by staff. Faced with a deluge of complaints over a minor line which had given the railway company so many recent difficulties, the Southern acquiesced. Consequently July 1925 saw the introduction of a fourth return passenger train, together with earlier and later departures. This gave better connections and met the needs of the milk traffic, but it meant that there was a six hour gap between trains in the middle of the day. Trains on the branch continued to be hauled by Adams' '02' tank engines, although it is probable other classes were used to shunt the sidings at both ends of the line on trip workings from Basingstoke and Alton respectively.

Somewhat surprisingly the Ministry of Transport did not seem in any hurry to inspect the line. Eighteen months elapsed before re-inspection was carried out. This eventually occurred on 4th February 1926 when Major Hall, rather than the usual Colonel Pringle, visited the line. His report is reproduced below;

> "I have the honour to report for the information of the Minister of Transport that I made an inspection on the 4th instant of the Basingstoke and Alton Light Railway section of the Southern Railway. This light railway, which was dismantled during the War, was re-opened for traffic in the early autumn of 1924. It is single line throughout and runs between Basingstoke, on the Company's main line, and Butts Junction where the double line from Alton, the single lines to Fareham and Medstead, and the Basingstoke Light Railway single line meet.
>
> Relaying has been done with second-hand 87 lb. rails of the old South Western section with sleepers of the usual dimensions, which are for the most part new. The chair fastenings are spikes and hollow trenails, and flint gravel ballasting has been used throughout.
>
> Generally speaking the line has been reinstated and is worked very much as it was before being dismantled, with the exception that there is now only one electric tablet section, namely between Basingstoke and Butts Junction, the intermediate tablet section and passing place at Herriard having been abolished.
>
> The signals at Basingstoke, giving access to and from the main lines, have been reinstated.
>
> There are a number of public road level crossings on the line, some of which are equipped with gates and attended by gate keepers, others being of the open cattle-guard type. In the case of the latter the road warning boards lettered 'Beware of Trains'

have been reinstated where they had been taken down, or renovated where they had not been removed. Opportunity has been taken in some cases of improving the position of these notices.
In regard to the level crossings equipped with gates, these are in all cases except one arranged so as to close across the road when rail traffic is passing, and to close across the railway when the crossing is open to road traffic. The exception is that at Lasham station, where the main Basingstoke - Lasham road a first class thoroughfare, crosses the railway. In this case the gates are arranged to close across the railway when open for road traffic but to open inwards towards the railway and do not close across the road, when the crossing is open for rail traffic. No doubt the reason for this is the objection which was taken by the County Council when the line was originally constructed to the encroachment caused by the gate posts when the gates were hung so as to close across the road when rail traffic was passing. The result of this alteration is that by far the most important road crossing on the railway has the least effective protection. I understand that the Company in order to reduce working expenses of the line desire to dispense with the gates and crossing keepers at some or all of these level crossings, and an application to this effect may be expected. With the exception of this main road crossing at Lasham Station the other roads at present protected by gates appear to be of a comparatively unimportant character, and at least two of them have the appearance of being little used farm cart tracks.
In regard to the cattle - guard crossings, since the arrangements here are merely a replacement of those previously existing and inspected I do not suggest that any alteration need be called for, but there are advantages in a form of lettering on the road warning boards which indicates definitely the presence of a level crossing and the absence of gates such as has been adopted in other light railways recently opened for traffic by this or other Companies. The question of the illumination either direct or by reflected light of the road warning boards at these crossings may require attention in the future.
Owing to the abolition of the tablet station at Herriard, the loop facing points on the single line have been removed and the points leading to a siding connection in the station yard are now worked from a two-lever ground frame controlled by the electric tablet for the section, and the interlocking of this ground frame is correct.
A new siding for the British Petroleum Company has also been laid, since the line was re-opened, near the Basingstoke end of the line. This is worked from a two-lever ground frame, the levers controlling respectively the bolt lock and the points, which is released by the electric tablet for the section, and the interlocking is correct. The siding connection is situated on the portion of the single line where the gradient is 1 in 88 falling from Basingstoke and the connection is worked so that the engine is at the lower end.
I recommend that the alterations at Herriard and the works in connection with the British Petroleum company's siding at Basingstoke be approved."

Signed Major Hall.

The re-introduction of train services brought a spate of accidents in the first year of operation. On 20th November 1924 the 2.05 p.m. mixed train from Alton in charge of Driver J. Norman damaged the crossing gates at Bentworth and Lasham station. It was reported that the gates should have been placed across the road by Porter William King, but he failed to hear the warning whistle because he was "busy in the yard". Both men were reprimanded.

Lasham road level crossing with the local freight train hauled by '02' 0-4-4 tank engine No. 234, the same locomotive that re-opened services in August 1924. The lady crossing keeper was paid the princely sum of 10/6d a week for carrying out her duties.

A more serious incident took place in early 1925 at Lasham level crossing when the 3.25 p.m. from Basingstoke hit a car driven by a Mr. Fletcher. Although there was no damage to the train the car was extensively damaged with its driver lucky to escape with only a bruised leg. The gatewoman had protected the train by exhibiting a red flag and the car had almost halted. However it would appear the car driver thought he could beat the train across the crossing. In this he was seriously mistaken and he was totally blamed for the accident.

Another set of level crossing gates were demolished on 10th August 1925 when the 11.40 a.m. goods from Basingstoke smashed into Shalden Crossing. No damage was caused to the unknown locomotive. In the official report it was stated that the resident crossing keeper was normally notified by bell of an approaching train. On this occasion the train was running ahead of time and consequently the gates had not been placed across the road quickly enough. Visibility was poor due to heavy rain and because of this the driver did not see the crossing until he was 80 yards away. Due to the extenuating circumstances no blame was attached to any party although it was recommended that a telephone be installed at the crossing to improve

The typical branch line train of latter years. An '02' 0-4-4 tank engine and its solitary coach depicted on the light railway, sometime in the late 1920's. *(J.G. Adams)*

'02' class 0-4-4 tank engine No 221 entering Alton Park platform at the head of a Basingstoke bound train. It is likely to be the same train as above. *(J.G. Adams)*

The branch train for Basingstoke leaving Alton on 17th August 1929. Henry Casserley's classic view of '02' 0-4-4 tank engine No. 221, a regular performer at this time. *(H.C. Casserley)*

communication with the stations and signal boxes at either end of the line.

Finally, in this catalogue of disasters, at 11.20 a.m. on 30th September 1925 two oil tank wagons being specially worked from Basingstoke to the BP siding were derailed at the points leading to the siding. Rerailing was completed by 1.30 p.m. and the line cleared by 1.48 p.m. The passenger service, which did not run between 11 am and 3.30 pm, was not affected although the goods working which usually left Basingstoke at 11.40 a.m. was cancelled. The cause of the incident was said to be subsidence under the siding trap point which did not allow the siding points to close properly.

Restricted usage and the need to reduce overheads led the Southern Railway to commission a census of the number of road vehicles using each of the gated level crossings during the week commencing 6th February 1926. This was an attempt to see if it might be possible to remove the gates and substitute cattle grids thereby reducing the number of staff at each station, the assumption being that the train guard would issue and collect tickets. Nevertheless it was recognised that a single Grade 1 Porter would still be required at each of the stations to deal with the goods and parcels traffic. The savings, which were negligible, would have involved axing all the station junior porters and the two crossing keepers. Recognising the limited savings and potential dangers to daily operation the information was not acted upon.

The blow that was waiting to happen was delivered in 1927 with the introduction by the Venture Bus Company of a competing service. Basingstoke seems to have had its first bus services in 1920 and Ford motor vans and lorries were on general sale in 1921 for £225 and £280 respectively. This coupled with the release of war surplus equipment led to an explosion of motor traffic in the 1920s. At a local level the Herriard Estate invested in a steam lorry at this time. Certainly Walker recognised the threat when the closure of the line was being discussed in 1923. The extent of the growth of motor vehicles can be seen by examining the Ministry of Transport's records. They show that in 1921 there were 873,700 motor vehicles and 1,100,000 people with a driving licence. By 1927 the figures were 1,898,500 vehicles and 2,349,000 licences. Consequently by the end of the decade there was a network of bus services radiating from Basingstoke and Alton directly serving the villages. An additional problem for the railway was that the local carrier, who had previously operated on a limited basis in the area was now able to offer an advantageous service using motor lorries.

The results for the railway from this threat to its passenger and freight business were immediate. By 1929 Venture were operating three weekday trips between Basingstoke and Alton and two on Sundays - the latter a service the railway had never provided.

There were three extra trips on Wednesdays (Basingstoke market day) and a late bus on Saturday nights for would be revellers. Fares were competitive; a single from Basingstoke to Herriard cost 8d and to Alton it was 1/6d. In comparison the corresponding train fares were 1/- and 1/9d. Serving the village centres whilst the stations were some distance from the villages they purported to serve, it was a case of 'no competition'. Railway receipts dropped like the proverbial stone whilst Venture could claim that 39,913 passengers used its buses in 1931. With a falling response to its 1925 improved service, the Southern Railway re-introduced in September 1927 a timetable of three trains each way. With inconvenient timings and long gaps between trains it is not surprising the bus company captured the market so completely.

Nevertheless the railway was to prove its worth during the winter of 1927. The latter part of December saw some of the worst snow conditions ever encountered in north Hampshire. From 27th December onwards thick snow blocked a number of lines in the area with the Hurstbourne to Fullerton branch, the Mid Hants and the light railway being affected. Despite tremendous efforts by railway staff assisted by a locomotive it was initially impossible to even reach Thornycrofts from Basingstoke. The line remained closed for some days and was finally opened on 2nd January 1928 when it temporarily became the sole means of access to the area.

With economy the order of the day in November 1927 the Southern applied to remove the level crossing gates at Hackwood Lane, Cliddesden and substitute them with cattle guards. This was agreed to by the Basingstoke and Alton Light Railway Amendment Order of 1928. With two level crossings Cliddesden had always been difficult to work and it was hoped that one staff member could be dispensed with.

Considering all the correspondence on the subject over the years, it is ironic that the final level crossing accident on the line occurred on one of the crossings not referred to so far. It was at the Thornycroft site where passage across the railway had been given in 1919 when they acquired the field on the opposite side of the line to the main factory. Mr Heaney, an employee of the motor manufacturer was driving a new lorry over the crossing just as the train approached from Alton. Edward Griffith, in his booklet, states that the train was unusually heavy with six coaches, but looking at the traffic figures for the line, this fact must be doubted. Nevertheless the locomotive struck the lorry hard leaving Heaney just holding the steering wheel. Anecdotal evidence says that the locomotive later arrived at Basingstoke still with the vehicle's radiator adorning the buffers. Fortunately no one was injured in the incident which is known to have occurred during 1928.

This was not to be the only time a locomotive hit a lorry on the line that year, although the other incident was deliberately planned and carefully executed.

It so happened that Gainsborough Pictures wished to film the stage play 'The Wrecker', written by Bernard Merivale and Arnold Ridley. The latter became well know through his earlier writing of 'The Ghost Train' and his much later appearances as 'Godfrey' in 'Dad's Army'. The story was somewhat simple, depicting a crook who organises a series of train crashes in order to discredit the railway in favour of the rival buses, a theme that would be used again by subsequent film makers. The silent film was directed by Geza von Bolvary and produced by Michael Balcon who subsequently went on to produce numerous Ealing comedies, notably 'The Ladykillers' which also had a strong railway theme. The actors were Carlyle Blackwell, Benita Hume, Joseph Striker and Pauline Johnson.

Alton station looking towards Farnham and London. The photographer is standing on the 'up' main whilst the 'up' relief is this side of the island platform. A Mid Hants train is about to depart from the down main behind a Drummond 'T9' 4-4-0.

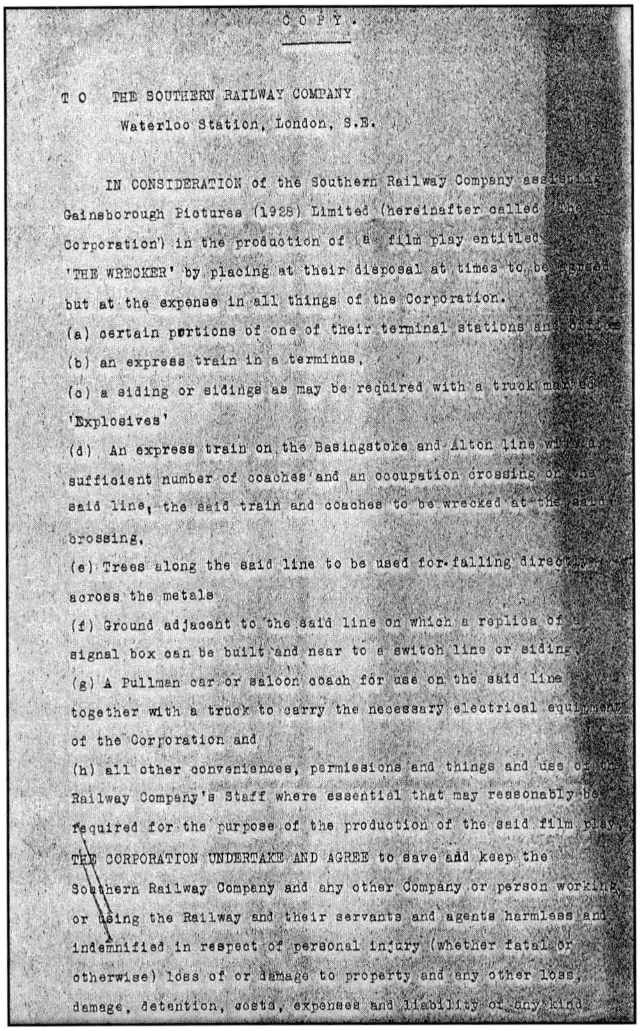

COPY.

TO THE SOUTHERN RAILWAY COMPANY
Waterloo Station, London, S.E.

IN CONSIDERATION of the Southern Railway Company assisting Gainsborough Pictures (1928) Limited (hereinafter called "The Corporation") in the production of a film play entitled 'THE WRECKER' by placing at their disposal at times to be agreed but at the expense in all things of the Corporation.

(a) certain portions of one of their terminal stations and offices
(b) an express train in a terminus,
(c) a siding or sidings as may be required with a truck marked 'Explosives'
(d) An express train on the Basingstoke and Alton line with a sufficient number of coaches and an occupation crossing on the said line, the said train and coaches to be wrecked at the said crossing,
(e) Trees along the said line to be used for falling directly across the metals
(f) Ground adjacent to the said line on which a replica of a signal box can be built and near to a switch line or siding,
(g) A Pullman car or saloon coach for use on the said line together with a truck to carry the necessary electrical equipment of the Corporation and
(h) all other conveniences, permissions and things and use of the Railway Company's Staff where essential that may reasonably be required for the purpose of the production of the said film play.

THE CORPORATION UNDERTAKE AND AGREE to save and keep the Southern Railway Company and any other Company or person working or using the Railway and their servants and agents harmless and indemnified in respect of personal injury (whether fatal or otherwise) loss of or damage to property and any other loss, damage, detention, costs, expenses and liability of any kind

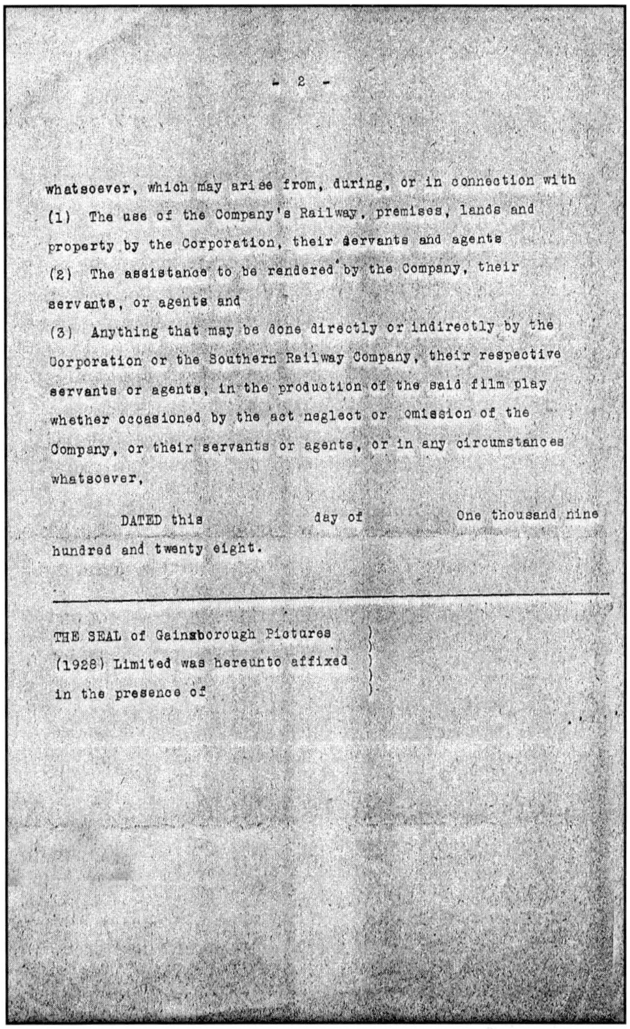

\- 2 -

whatsoever, which may arise from, during, or in connection with
(1) The use of the Company's Railway, premises, lands and property by the Corporation, their servants and agents
(2) The assistance to be rendered by the Company, their servants, or agents and
(3) Anything that may be done directly or indirectly by the Corporation or the Southern Railway Company, their respective servants or agents, in the production of the said film play whether occasioned by the act neglect or omission of the Company, or their servants or agents, or in any circumstances whatsoever,

DATED this day of One thousand nine hundred and twenty eight.

THE SEAL of Gainsborough Pictures (1928) Limited was hereunto affixed in the presence of

Contract between the Southern Railway and Gainsborough Pictures (1928) Limited for the filming of "The Wrecker".

Earlier in the year the film company had approached the Southern, seeking permission to use one of their lines. The railway company responded by offering the Basingstoke and Alton line, ideally suited with its light traffic, open landscapes and limited housing nearby. The most spectacular incident took place on Sunday 19th August 1928 at Salter Hatch crossing near Lasham when a complete train hit a Foden steam lorry which had been placed across the line. The Southern, willing to assist the film company sold them an obsolete rake of six coaches and a locomotive especially for the occasion. The coaches were notable for being the first bogie vehicles built by the former South Eastern and Chatham Railway whilst the locomotive was a James Stirling 'F1' class 4-4-0 originally built by the South Eastern Railway in 1889 as a class 'F'. As designed the locomotives were domeless with a small severe cab, and with their 7 ft driving wheels they were intended for use on the main express duties. The locomotive in question No. A148 was rebuilt with a domed boiler and a conventional cab by Harry Wainwright in 1903 and re-classed as 'F1'. The film company repainted the locomotive in grey with the lettering 'United Coast Lines' on the tender.

The preliminary work which included undermining and loosening the track had to be done after the passage of the last train on the Saturday. Meanwhile dummy haystacks and trees were built to hide the cameras. On the day in question two breakdown cranes were put into position - probably at Basingstoke and Alton, so as to be ready to clear the debris, whilst at 6.30 a.m. a 'special corridor dining train' from Waterloo arrived with the chief officials of the railway; later some two hundred film crew arrived by car. North Hampshire had never seen anything like it before (or since!) and police had to be drafted in to keep back the public. By the early afternoon everything was ready and the Guildford crew of driver James Brown and fireman Goodright set the train rolling and then jumped clear. Accelerating down the 1 in 50 gradient the locomotive demolished the lorry and then rolled over on to its left hand side. To complete the action the coaches were then covered with petrol and ignited. With the day's events complete the special train left Herriard at 6.55 p.m., with driver Hudson at the controls. The task of

The scene a second before the deliberately staged crash at Salter Hatch crossing, Lasham. The six coach train is accelerating down the 1 in 50 gradient and is about to hit the steam lorry. The track in the foreground has been undermined ensuring that the train will de-rail after the collision.

A moment later and the dramatic effect of the impact is captured in this shot taken from a specially constructed camera tower. Behind the locomotive is another camera 'hide' disguised as a haystack.

A clear view of the crash with the lines of the locomotive being obliterated by the ballast in the lorry. The two dummies can be seen to be fully in charge. The event was also covered extensively in the national press the following day.

Steam escapes from the derailed locomotive which has fallen on to its left side. The two camera towers are in the foreground and the camera 'hide' is in the field beyond the line. Considering the danger associated with the filming of this scene the presence of the two individuals on the track by the level crossing is most surprising.

Boys will always want to be near the action. The public survey the wrecked train now that the boiler has been made safe. Although the cab sheet has been ripped from the boiler, the damage to the locomotive was not extensive.

"Passengers" are being helped from the wreckage. Although derailed the coaching stock survived the impact well.

The final part of the drama is enacted with the rolling stock falling victim to fire. The liberal application of petrol ensured the conflagration would be worth watching by cinema viewers. Today's railway preservationists can only wince at the wanton destruction of such historically important items.

removing the wreckage was quickly done, being completed by 10 a.m. the following day. However normal working was only resumed in the afternoon with the passage of the 3.25 p.m. from Basingstoke - the last train of the day. Apart from giving the Southern Railway some excellent publicity and allowing them to dispose of unwanted rolling stock it also gave them some welcome income. Without a doubt, the £6,000 to £7,000 earned was the most the line was ever to achieve in its working life. The film company, keen to maximise its investment in a costly episode, subsequently used the footage in the 1936 film 'Seven Sinners'.

After this excitement, the line settled down to its normal quiet routine. Milk traffic in churns from Bentworth and Lasham and Herriard still continued to be an important traffic, being conveyed to Waterloo via Alton on the first train and via Basingstoke on the last. Freshness could be a problem with the journey taking up to 7½ hours. Revenue continued to fall and in late 1930 consideration was given to the line's closure. A letter dated 20th November comments that "the matter must remain in abeyance until 1934 when it may be necessary to examine the question as to whether the line shall be closed altogether" - the reason for the delay being that 1934 would then be the end of the ten year period which the Southern Railway had assured the House of Lords that services would be maintained. However Walker, mindful of the losses on several branch lines at a time of international economic uncertainty set about the closure of the branch in 1932. Consequently on 11th April 1932 the Southern Railway wrote the following letter to all the local authorities and landowners:-

> "I have to inform you that in consequence of the paucity of traffic on the Basingstoke and Alton Light Railway the earnings compare very unfavourably with the expenses incurred in the maintenance and working of the railway, and it has, therefore been decided to close the line for passenger traffic at the

end of the ensuing summer season, it being felt that the needs of the district can be adequately provided for by the means of road transport."

Only Major Jervoise replied and even then he agreed that there was no other course of action, although he did criticise the railway company for the services it had offered. He said that "no scheme of running trains was devised to suit either those passengers who desired to spend a day in London or for the convenience of the village women who desired to go into Basingstoke or Alton to spend from an hour to two hours shopping, and it was for this reason that the Venture Bus Company started their service with the success so far as the latter class of passengers are concerned". It can be seen that the meaning of political correctness had yet to be learnt and that the English class system was alive and well in north Hampshire. From the local authorities who had campaigned so hard throughout the line's history, not a word was heard. Likewise, Col. the Rt. Hon. W.G. Nicholson MP who had swayed the Lords in 1923, could not be bothered to reply. The motor vehicle had created indifference and apathy to a transport system that was now outmoded.

On 14th June 1932 the following internal memorandum was sent to Walker:-

"It will be seen that the majority of those who were in opposition to our proposal to abandon the line in 1923 have taken no notice of the information contained in the above mentioned letter of April 11th.

The Alton Rural District Council asked whether the railway will continue to be worked for goods traffic and they have been assured that a service will be afforded over week-days to and from Cliddesden, Herriard, and Bentworth via Basingstoke, conditionally on traffic offering.

The comment made by Major Jervoise is true in regard to the train service given when the line was re-opened in August 1924, but in the summer of 1925, an improved passenger service, which gave a morning train to London and some facilities for shopping in Alton and Basingstoke, was provided at considerable additional expense.

This improved train service was continued for two years, with very disappointing results and the first morning train in each direction was withdrawn in September 1927, there having been no indication that the passenger traffic could be held even with an extravagant train service. The passenger traffic has continued to diminish.

In addition to the traffic to and from the branch stations there is a small through traffic between Basingstoke and Alton now amounting to 100 to 130 passengers per month in each direction.

In the heaviest month last year (August) the average number of passengers per train was five only.

The estimated earnings of the branch line for passenger traffic for 1931 were only £350 and the contributive value is small, as very few tickets are issued to or from stations beyond Basingstoke and Alton respectively.

The estimated branch line earnings for parcels and miscellaneous traffic for 1931 were £100, principally accruing from milk traffic despatched from Herriard. Arrangements can be made to collect the morning despatch of milk (about 20 churns) from Herriard Station by lorry from Basingstoke at a cost of 5/3d per day, and the two or three churns forwarded in the afternoon can be collected by a lorry service which has operated for some time past to collect milk in bottles from a farm in the neighbourhood. The principle milk sender, Mrs. Jervoise, has asked what arrangements will be made for the conveyance of her traffic when the passenger service is withdrawn and has been assured that a suitable lorry service will be provided.

With the discontinuance of the passenger service there would be no necessity to use the line between Bentworth and Butts Junction, as the whole of the goods traffic could be conveniently worked via Basingstoke, and it is therefore proposed to convert the line to a goods siding and remove the permanent way between Bentworth and Butts Junction.

The cost of the work involved in this proposal is estimated by the Engineer to be £1,860, but the value of the material to be recovered is £3,657.

The adoption of the proposal would result in a net saving of £1,668 per annum as shown hereunder:-

Loss or increased expense per annum		**Gain per annum**	
Passenger traffic	£350	Saving of one pair of enginemen and 504 train miles per week	£860
Cartage of milk from Herriard at 5/3d per day	82	Withdrawal of one passenger guard, one junior porter and two female crossing keepers	300
		Saving in annual maintenance	760
		Saving in annual commuted sum for ultimate renewal of displaced works.	180
	£432		£2100

Net saving £1,668 per annum.

You may deem it advisable to obtain powers for the abandonment of the portion of the light railway between Bentworth station and the Cripple's Home platform at Butts Junction before the permanent way is removed, but in the meantime I recommend the discontinuance of the passenger train service as from Monday 12th September, together with the provision of the necessary safeguards for working the portion of the line between Basingstoke and Bentworth as a siding.

As the passenger traffic is so small, I do not think any special arrangements need be made as to road conveyance, the omnibus service operated by Venture Ltd...... affording the necessary facilities."

On 21st July Walker reported the unfavourable situation to the Traffic and Continental Committee who authorised the closure from Monday 12th September - two years less than the ten year 'guarantee'.

The last train left Alton on Saturday 10th September 1932 at 7.30 p.m. with only one passenger on board. This was Mrs. Eliza Bowman who had the distinction of having travelled on the first train back in 1901. In an era before railway enthusiasts and local people clamoured around "last trains" the 'occasion' was quite uneventful. In spite of the closure the inefficiency of the massive Southern Railway organisation meant that cheap tickets over the light railway continued to be advertised as available until almost the end of

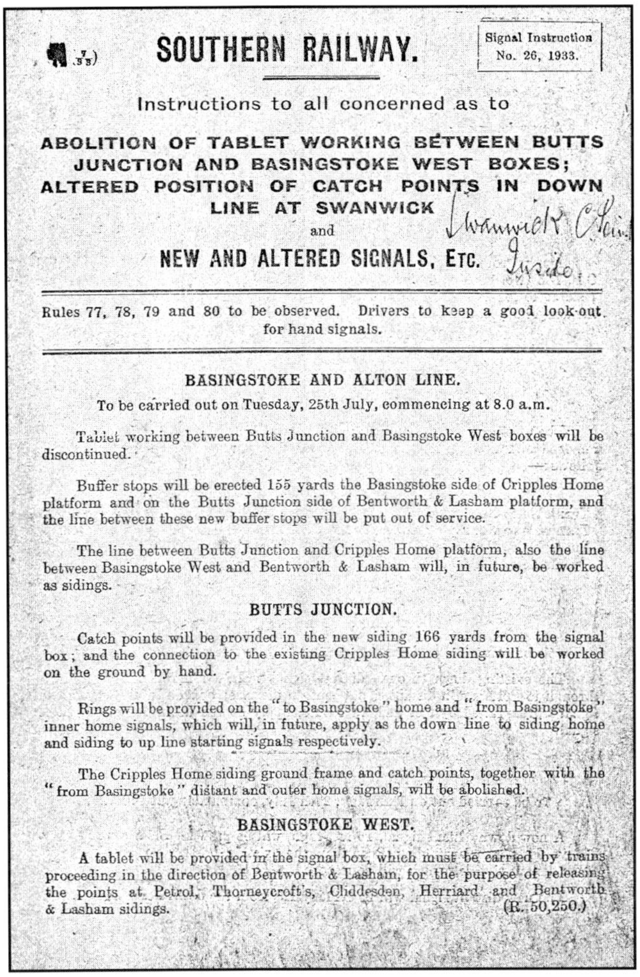

SOUTHERN RAILWAY.

Signal Instruction No. 26, 1933.

Instructions to all concerned as to

ABOLITION OF TABLET WORKING BETWEEN BUTTS JUNCTION AND BASINGSTOKE WEST BOXES; ALTERED POSITION OF CATCH POINTS IN DOWN LINE AT SWANWICK

and

NEW AND ALTERED SIGNALS, ETC.

Rules 77, 78, 79 and 80 to be observed. Drivers to keep a good look-out for hand signals.

BASINGSTOKE AND ALTON LINE.

To be carried out on Tuesday, 25th July, commencing at 8.0 a.m.

Tablet working between Butts Junction and Basingstoke West boxes will be discontinued.

Buffer stops will be erected 155 yards the Basingstoke side of Cripples Home platform and on the Butts Junction side of Bentworth & Lasham platform, and the line between these new buffer stops will be put out of service.

The line between Butts Junction and Cripples Home platform, also the line between Basingstoke West and Bentworth & Lasham will, in future, be worked as sidings.

BUTTS JUNCTION.

Catch points will be provided in the new siding 166 yards from the signal box; and the connection to the existing Cripples Home siding will be worked on the ground by hand.

Rings will be provided on the "to Basingstoke" home and "from Basingstoke" inner home signals, which will, in future, apply as the down line to siding home and siding to up line starting signals respectively.

The Cripples Home siding ground frame and catch points, together with the "from Basingstoke" distant and outer home signals, will be abolished.

BASINGSTOKE WEST.

A tablet will be provided in the signal box, which must be carried by trains proceeding in the direction of Bentworth & Lasham, for the purpose of releasing the points at Petrol, Thorneycroft's, Cliddesden, Herriard and Bentworth & Lasham sidings. (R. 50,250.)

Signal Instruction No. 26, 1933 giving the signalling changes introduced after the closure of services between Treloar's and Bentworth and Lasham.

September. The previous limited response by the travelling public meant few, if any, potential customers were inconvenienced by this mistake.

As promised the weekday goods service was continued, the only difference being it now operated from Basingstoke to Bentworth and Lasham only. On Mondays to Fridays the train left Basingstoke at 1.30 p.m. arriving at Bentworth and Lasham at 2.37 p.m. It returned 20 minutes later arriving at Basingstoke at 3.49 p.m. On Saturdays the initial departure was at 11.55 a.m. with a return time of 2.05 p.m. In conjunction with this service loaded and empty milk churns were conveyed to and from Herriard station by motor vehicle.

Having previously been bitten the Southern Railway were in no hurry to lift the track south of Bentworth and Lasham and waited until the necessary parliamentary powers for abandonment had been secured. These were obtained in 1933 and only then were buffer stops placed on the Basingstoke side of Bentworth and Lasham level crossing with another set 700 feet beyond the platform at Alton Park. An assurance was given to the hospital authorities that their private siding facility would be maintained as would a service to the hospital platform as required.

At this time the stub remains of the railway were still operated under the tablet system which was outmoded now that the line was severed. Despite this peculiarity, such working continued until 18th July 1933 when SR Signal Instruction No. 26 gave details of future modifications to be carried out to the points and signals on the line to enable the remaining section in use to be worked as a long siding from Basingstoke. This involved the conversion of the siding point to Messrs BP and Thornycroft's to hand operation, as well as those leading to the sidings at the three stations. Further work associated with these changes was carried out to the locking and signals at Basingstoke West signal box. All the above changes were put into effect from 21st August 1933. Special instructions were now issued to cover the working:-

> "When it is necessary for a goods train to run from Basingstoke to Bentworth and Lasham, or some intermediate point, the driver, also the guard or shunter, must be instructed from what point the train is to return. The train must be brought to a stand and the engine whistle sounded before passing over any of the level crossings and particular care must be exercised....during fog or falling snow."

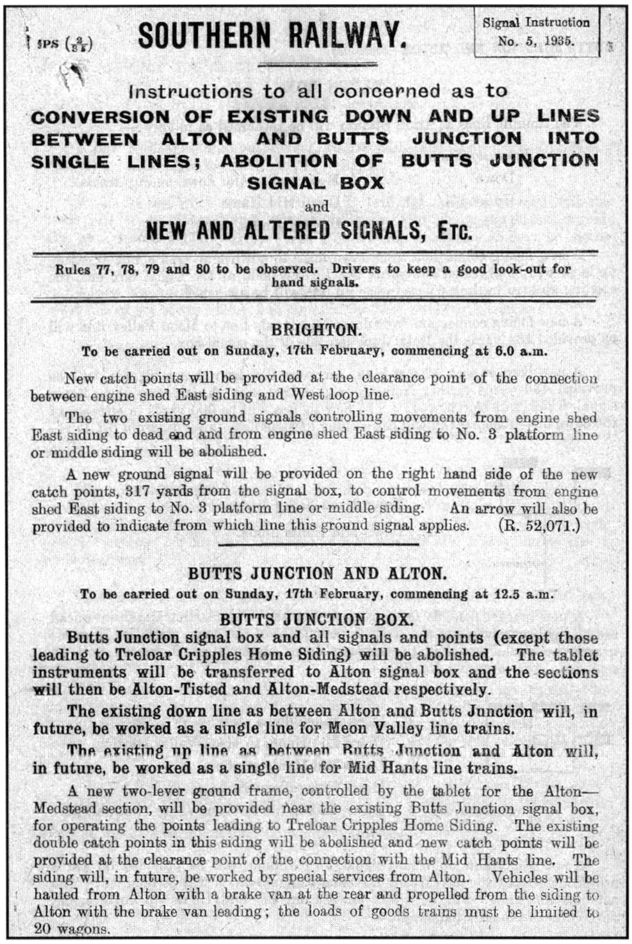

SOUTHERN RAILWAY.

Signal Instruction No. 5, 1935.

Instructions to all concerned as to

CONVERSION OF EXISTING DOWN AND UP LINES BETWEEN ALTON AND BUTTS JUNCTION INTO SINGLE LINES; ABOLITION OF BUTTS JUNCTION SIGNAL BOX

and

NEW AND ALTERED SIGNALS, ETC.

Rules 77, 78, 79 and 80 to be observed. Drivers to keep a good look-out for hand signals.

BRIGHTON.

To be carried out on Sunday, 17th February, commencing at 6.0 a.m.

New catch points will be provided at the clearance point of the connection between engine shed East siding and West loop line.

The two existing ground signals controlling movements from engine shed East siding to dead end and from engine shed East siding to No. 3 platform line or middle siding will be abolished.

A new ground signal will be provided on the right hand side of the new catch points, 317 yards from the signal box, to control movements from engine shed East siding to No. 3 platform line or middle siding. An arrow will also be provided to indicate from which line this ground signal applies. (R. 52,071.)

BUTTS JUNCTION AND ALTON.

To be carried out on Sunday, 17th February, commencing at 12.5 a.m.

BUTTS JUNCTION BOX.

Butts Junction signal box and all signals and points (except those leading to Treloar Cripples Home Siding) will be abolished. The tablet instruments will be transferred to Alton signal box and the sections will then be Alton-Tisted and Alton-Medstead respectively.

The existing down line as between Alton and Butts Junction will, in future, be worked as a single line for Meon Valley line trains.

The existing up line as between Butts Junction and Alton will, in future, be worked as a single line for Mid Hants line trains.

A new two-lever ground frame, controlled by the tablet for the Alton—Medstead section, will be provided near the existing Butts Junction signal box, for operating the points leading to Treloar Cripples Home Siding. The existing double catch points in this siding will be abolished and new catch points will be provided at the clearance point of the connection with the Mid Hants line. The siding will, in future, be worked by special services from Alton. Vehicles will be hauled from Alton with a brake van at the rear and propelled from the siding to Alton with the brake van leading; the loads of goods trains must be limited to 20 wagons.

Signal Instruction No. 5, 1935 detailing the closure of Butts Junction signal box.

Butts Junction looking west in 1953 clearly showing the divergence of all three lines. To the right is the former signal box now relegated in status to that of permanent way hut. *(Denis Cullum).*

It would appear no form of signalling was in use although it could well be that an unofficial 'wooden staff' or similar was retained as a reminder to the signalman in the busy Basingstoke West box.

Whilst these changes were being implemented the track was lifted between Bentworth and Lasham and Alton Park with the ballast recovered being used to make up Kings Road, Alton. Meanwhile the Southern Railway dusted off the 1922 plans to close Butts Junction signal box and this took effect from 17th February 1935. As previously planned the former double track to Alton station was converted into two parallel single lines connecting at the station. Access to the brewery siding and Treloar's was maintained with a release secured from the Alton to Medstead tablet section. In later years this was altered to Tyer's electric token working. Butts Junction signal box was converted into a permanent way hut by the removal of the operating floor and the placement of a pitched roof on what was once the locking room.

In spite of these economies, rather than because of them, the traffic on the line continued to be limited. Eighteen months after the "guarantee" expired Walker recommended full closure to the Board as the traffic "is very small" with "little increase in the future.... anticipated". His view was accepted and accordingly on 31st March 1936 the closure was confirmed in a further memorandum.

"On and from 1st June next, Cliddesden, Herriard and Bentworth and Lasham stations will be closed to traffic and all services will be withdrawn from the section of the Basingstoke and Alton Light Railway between Thornycroft's siding, Basingstoke and Bentworth and Lasham.
Delivery and collection of general goods traffic, also parcels in the area served by the light railway will be effected from and to Basingstoke by our cartage and arrangements for the delivery of coal and coke traffic from Basingstoke are being made with a local haulier.
Traffic to and from Thornycroft's siding will continue to be dealt with as at present, a short length of line being retained for use as a shunting neck in connection with the working of the siding."

News of the line's impending closure appeared in the local press in mid-May. There were no protests, no indignant letters and no demands for retention - the railway had become an anachronism. In this the line merely predated the pattern that became commonplace in the post World War II years.

Devoid of fuss or ceremony the final goods train ran on 31st May 1936. This last service also had the task of collecting salvageable stores from the stations and was in charge of driver William Hudson and fireman Stanley R. Barton. Hudson it will be recalled had also been involved with the first trains in 1901 and 1924 as well as the special in conjunction with the filming of 'The Wrecker' - surely a unique qualification.

Neglected and rusting, the railway awaited its end

which began in 1937 with the lifting of the rails from Bentworth and Lasham and on towards Basingstoke. The line which had cost over £25,000 to lay was going to produce a net value of only £3,000 to the railway company. However fate was about to change its short term destiny. Whilst track lifting was in progress and had got as far as Herriard, Gainsborough Pictures again approached the Southern Railway for the use of one of their branch lines to film the picture "Oh, Mr. Porter!" Starring the then popular comedy trio, Will Hay, Moore Marriott and Graham Moffatt the story is set on the 'Southern Railway of Northern Ireland' and concerns an incompetent Station Master and his staff who defeat a gang of gun runners. The film company accepted the offer to use the light railway and Cliddesden was transformed into 'Buggleskelly'. The station building was encased in timber and a signal box and signal built adjacent to the level crossing at the Basingstoke end of the station. The sidings were used for shunting whilst the running line was used for numerous sequences. Prominent were the boards restricting trains to 10 m.p.h. whilst a false tunnel was constructed in a cutting.

Three locomotives were used in the film, two being supplied by the Southern with the third coming from the Kent and East Sussex Railway. Adams' 'X6' class 4-4-0 number 657 was used as the motive power for the express train. Built in December 1895 at the LSWR's Nine Elms works the locomotive was the first of a class of ten designed for main line work. By the time of filming half of the class had been scrapped and the remainder were used on secondary duties. The choice of 657 was probably decided by the fact that Basingstoke was its home shed and that it was nearly life expired - three years after filming it was scrapped. The excursion train was hauled by Adams' '395' class

Will Hay, Moore Marriott and Graham Moffatt (Right to left) enter the "tunnel". Behind is "Gladstone", the KESR 2-4-0 tank engine.

Will Hay, Graham Moffatt and Moore Marriott on the footplate of KESR 2-4-0 'Northiam', alias "Gladstone". The maker's plate is quite clear whilst the Rother Valley Railway lining is showing distinct signs of ageing.

Cliddesden station, alias "Buggleskelly". The SR post and wire fencing is clearly shown whilst in the background the sidings are still in situ. The postman's quote "you're wasting your time" could have equally applied to the light railway's promoters and the supporters for re-instatement.

Will Hay's sleep pattern is disturbed at Buggleskelly. The carriage was a studio mock-up but the cow was real enough being owned by Jack Hooper whose grandson still lives in Cliddesden.

Graham Moffatt with a 'gun runner'. This was part of the finale and was filmed on the move although this shot appears posed as the coach is standing at a station platform.

"Gladstone" demolishes a level crossing. Graham Moffatt is precariously clinging to the carriage roof armed with a shovel.

0-6-0 number 3509. Built at the works of Neilson and Co. in December 1885 it had their works number 3389. The class was designed for main line goods work but could also be seen on secondary passenger turns. At the time of filming the locomotive was shedded at Eastleigh. The locomotive had a long life, being finally condemned as British Railways 30581 in March 1953. Minor alterations were carried out to both locomotives. The cab side number plates were amended by the painting of the wording "of Northern Ireland" below the name "Southern Railway" and they were given shorter ex-LBSCR chimneys.

From a locomotive viewpoint, the star was the Kent and East Sussex engine. They supplied a 2-4-0 tank engine with outside cylinders which had been built for the Rother Valley Railway (a precursor of the KESR) in 1899 by R.W. Hawthorne Leslie of Newcastle-upon-Tyne, their works number 2421. Named 'Northiam' the locomotive still sported its original Rother Valley livery of dark blue lined out with white, black and vermilion. However at the time of filming the original colouring had weathered considerably and was a somewhat grimy green/black. For the film the rear of the cab was cut away and a tall extension to the chimney with a serrated top was fitted deliberately giving the engine a humorous and ancient appearance. It also meant that the actors could be seen to better effect. The nameplate was removed and replaced by another, "Gladstone", although the worksplate was left on the bunker. The locomotive was seen at Basingstoke on 13th June 1937 having travelled via Tonbridge, Redhill, Guildford and Reading with a crew from two of the Col. Stephen's empire of light railways - the Kent & East Sussex and the East Kent. When filming was complete, 'Northiam' returned by the same route back to Kent in early August. It lasted only a few more years and was scrapped in 1941.

The arrival of the film crew in mid June 1937 caused much excitement with the local paper reporting that the peaceful slumber of the village (Cliddesden) was aroused. In all the film unit numbered about fifty - a considerable influx into such a small community. To record the action shots a platform was affixed to the side of No. 657 which was well out of loading gauge and projected over the earthworks. However there are no records of any accidents. Filming took about two months with the final scenes being made in Basingstoke West yard. "Gladstone" had to weave its way through the sidings where the two Southern locomotives were strategically placed. The shot was used extensively with the scenes being taken from differing angles. In addition the observant viewer of the film will also catch a brief glimpse of Adams' "02" class 0-4-4 number 200 stationary in the yard and a departing express from Basingstoke.

The Southern allowed the film company the use of other parts of the old LSWR main line. The title sequencies were filmed between Basingstoke and Micheldever where good use was made of Popham tunnels and the dramatic entrances through deep cuttings. The sharp bend from Northam to Southampton and the entrance to Southampton Central feature twice whilst the entrance to "Belfast" shows sections of third rail electrified track and was filmed between Clapham Junction and Nine Elms. A blunder(?) which was never rectified was the reversal of the film when the titles were being prepared. Consequently many of the main line shots give the impression of a train travelling on the right hand track, a fact that could be forgiven in a Will Hay comedy. By late summer, all was finished and as the 'Southern Railway Magazine' commented "Cliddesden's challenge to Hollywood is no more". At the same time the front cover of the same magazine had a portrait of Sir Herbert Walker, announcing his retirement as General Manager.

With filming complete the track lifting recommenced and by the end of 1937 it was probably complete. Meanwhile traffic continued to the various sidings on the former route, Treloar's for example handling up to 3,000 tons of coal per annum by the late 1930's. Thornycrofts too were expanding and in May 1949 an additional siding on the opposite side of the line was provided at an estimated cost of £2,000.

By the 1960's British Railways was finding that private sidings, often with single wagon traffic, were proving to be a financial liability. This was the era of the 'Beeching cuts' when the railway management ruthlessly cut unprofitable operations. The first to go was the BP siding which was removed in October 1964. This was followed in 1967 when British Railways served notice to terminate the private siding agreements and traffic over the two stub ends of the branch. A stop block was erected on the Basingstoke side of the waterworks level crossing and from then on traffic on the Basingstoke and Alton Light Railway ceased totally. As Moore Marriott said in "Oh, Mr. Porter!", "the next train's gone!"

Following the lifting of the track, the Southern Railway progressively sold off the land that had been compulsorily acquired. In 1946, the Earl of Portsmouth bought 13³/₄ acres at Cliddesden for £1,860, his predecessor having sold 18³/₄ acres for £1,170 in 1900. Similarly in 1941 the Herriard Estate purchased the 43 acres it had sold in 1900 at a price of £5,200.19s, for £2,250. Credit for the best deal of all must go to Mr. & Mrs. Gaywood who bought just over six acres at Shalden in July 1945 for a mere £20. The capital invested so readily in the late 19th century was certainly not re-couped when the assets were sold. Like so many rural railways it was a poor investment in every sense.

Coal for Treloar's in 1957 with '700' class 0-6-0 No. 30697 having just turned away from the Mid Hants line on to the truncated remains of the light railway.

Alton Park in latter years. The turnout in the foreground led back towards the coal siding.

Treloar's siding showing the winch which drew wagons into and out of the hospital premises.

'2-BIL' unit No. 2054 awaiting departure from Alton towards Farnham and London in the early 1960's. At this time the former up relief was used for Mid Hants services only, hence the fact it was non electrified. *(Sean Bolan)*

The branch line finally severed from the Mid Hants line after the cessation of coal traffic to Treloar's.

One of the last passenger workings to Alton Park. The 'Railway Enthusiasts Club' of Farnborough 'North Hampshire' tour of 15 October 1960. At the head was M7 No. 30028. A number of locations were visited in the area with push-pull set No. 1 which had seen regular use on the nearby Meon Valley line some years previously. *(David Lawrence)*

The 'Rambling Rose' railtour special at Thornycroft's works on 23 March 1963. This was probably the very last passenger working on the line.

Thornycroft's siding in 1966 looking towards Basingstoke. That to the right is part of the original facilities whilst to the left is the new siding added in 1949. The road crossing was approved in 1919 and it was here that the branch train struck a lorry in 1928. In the distance it is just possible to discern some wagons standing on the main line; the descent to the factory from the junction is quite evident.

The parting of the ways - this is not a branch train! 'Merchant Navy' class 4-6-2 No. 35011 'General Steam Navigation' at the head of the down Bournemouth Belle west of Basingstoke on 29th March 1955. The train is travelling on the down fast line and to the right are the down slow and the former Alton branch curving away towards Thornycroft's. *(Philip J. Kelley)*

Cliddesden village pond with residents c.1905. It was from communities such as this that the railway was intended to cover its operating costs. By 1929 Venture were running an intensive service to the village terminating at this point.

CHAPTER 8
THE STATIONS – CLIDDESDEN

Approximately two miles south of Basingstoke, Cliddesden village is best described by its description in the 1915 Kelly's Directory - "...a pleasant village....". Pleasant indeed but with little perhaps to commend it to history although it does possess a parish church built of stone in the Norman and Perpendicular styles. It was restored in 1868 at a cost of £500 by the 5th Earl of Portsmouth, the Rev. J. Bryan, rector from 1841-87 and, of course, the inhabitants of the village.

A variety of tradesmen both lived and operated in the village, and included a blacksmith, grocer, cycle dealer and wheelwright. Indeed it could be said that the parish was relatively self sufficient with its own post office, elementary school, public house (the Jolly Farmer) and even a registrar of births and deaths. However, one noticeable omission was that of a doctor.

Located on the Basingstoke to Alresford road the various styles of houses clustered around the village pond. During the Victorian years the population of the parish remained static; between 1831 and 1921 it changed from 329 to 316! At no time in that period did it exceed 334 or drop below 306; when the railway opened in 1901 the number stood at 321. This stability reflected the relatively unchanging nature of the local economy with agriculture being the major employer. Much of the land was arable with crops of wheat, barley and roots, although pasture increased in importance over the years as corn prices fell. Dominating everything was Hackwood Park the ancestral home of the Earl of Portsmouth. The house was designed by Inigo Jones but the dominating feature of the park was an impressive aisle of beeches known as the Cathedral.

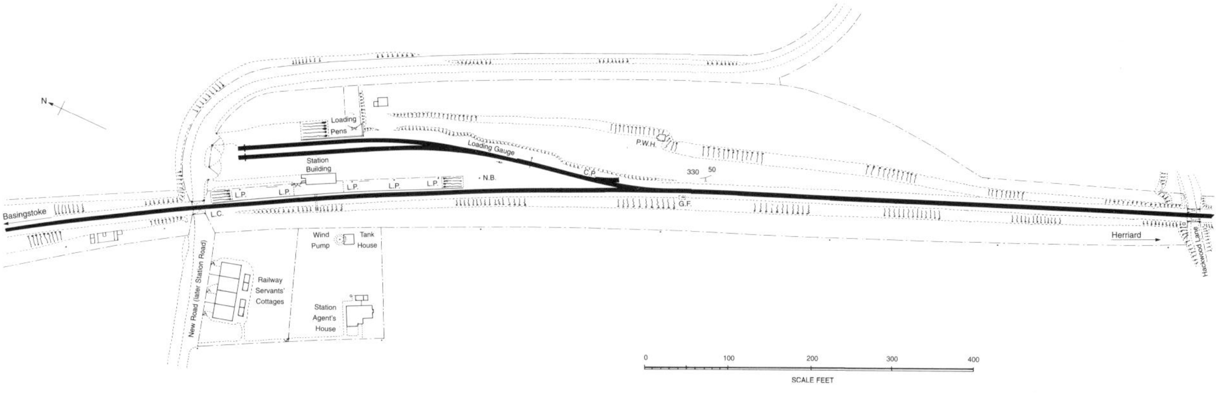

CLIDDESDEN - 1911

It was into this unchanging environment that the railway entered. The station was located about $\frac{1}{2}$ mile from the village (not $1\frac{1}{2}$ miles as the newspaper report on the opening stated) on an exposed site approximately 400 feet above sea level. It consisted of a single platform on the east side of the line behind which were two sidings forming the goods yard. There was no goods shed or crane but one of the sidings was served by a loading ramp and also faced some livestock pens. At the exit of the yard was a loading gauge - essential when potential out of gauge loads such as hay or straw were being handled. As required for safe working a catch point protected the entry on to the single running line. The turnout from the running line was controlled by a two lever open air ground frame which was itself released by the single line tablet.

The station building itself was small and had limited accommodation. Nevertheless it was all that was required for the restricted traffic that was anticipated and in fact dealt with. Constructed of corrugated iron under a roof of the same material, it measured 35 feet x 12ft. 6in. (scaled) and was divided internally into three equal rooms. The first of these was the booking office, then came a booking hall/waiting room with the third being used as a parcels/store room. All had a match board panelled interior. A single toilet cubicle was provided in a small extension off the public area. Up to 1917 the station also acted as a telegraph office, the only one on the light railway, and this facility was available to the public during the hours the railway was open. The platform was 250 feet long with a depth of 12 feet and a height of 3 feet. At both ends were ramps of 1 in 8. It had a

The neat and tidy appearance of the station is apparent in this Edwardian view which faces Alton. Mr. Bushnell seems to have attracted an army of young supporters.

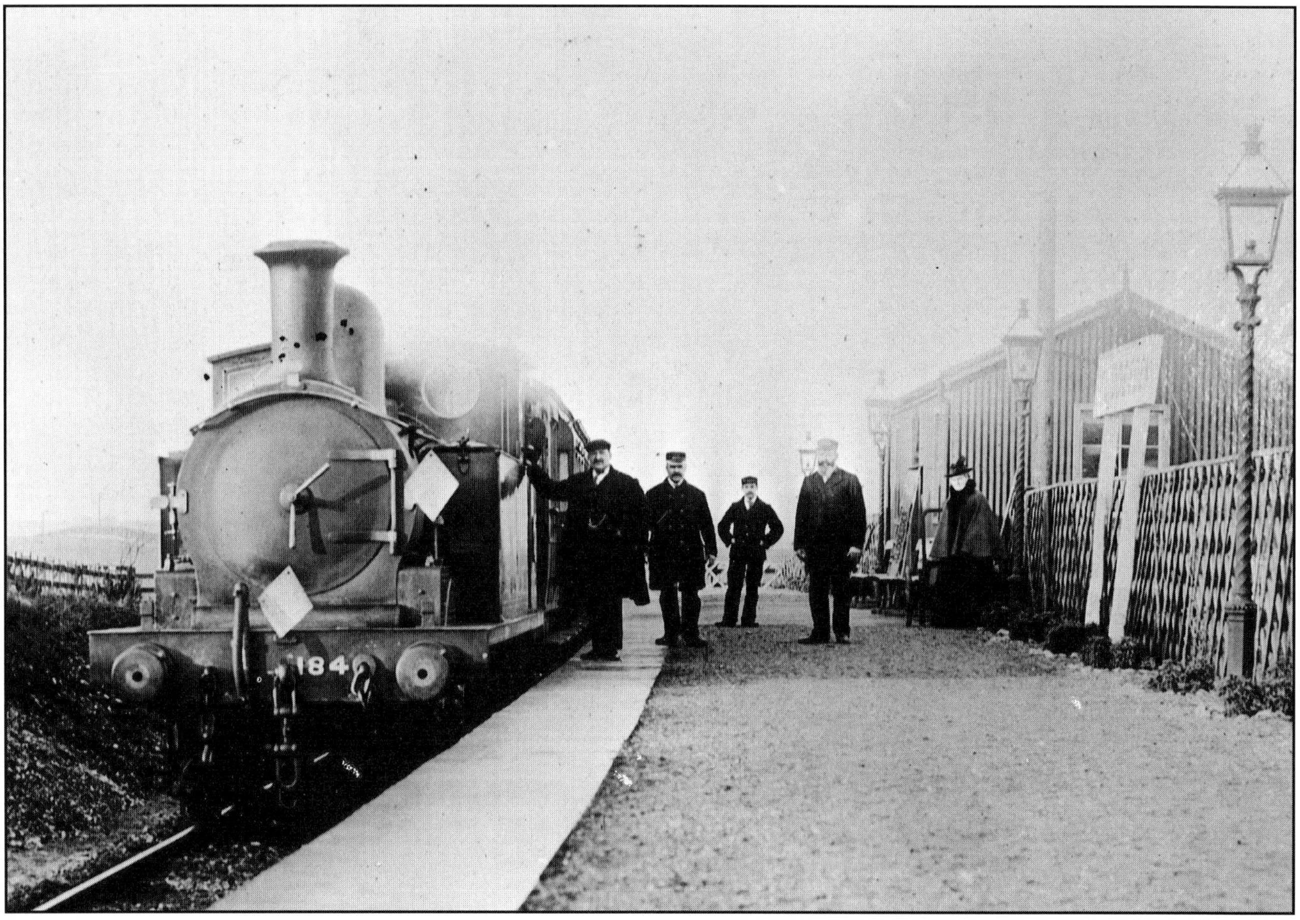

Early days at Cliddesden with an Alton bound train paused at the platform behind '02' 0-4-4 No. 184. Mr Bushnell is the gentleman on the right whilst the 'lad porter' is next to him with his arms on his hips. The photograph shows details of the station including the LSWR 'candy twist' lamps and the trellis fence which was a feature of all the stations on the line. No celebrations marked the opening of the line in 1901 although more were undertaken in 1924 when the line re-opened.

tarmacadam surface in the vicinity of the station building, the remainder being gravel. Five oil lamps of standard LSWR 'candy twist' type illuminated the platform, although an early photograph shows a different more basic type in situ. Behind the platform was a trellis type of wooden fence. This was replaced by a standard Southern Railway concrete post and wire fence, probably in 1924 when the line re-opened. To give the station a measure of protection a row of horse chestnut trees was planted behind the platform.

Pedestrian and vehicular access to the site was by the purpose built road demanded by the Light Railway Commissioners and called, not surprisingly New Road (later Station Road). Branching from the well established Hackwood Lane on the edge of the village, this was the road access agreed upon as part of the compromise reached over the station location disagreement. It was built by Firbank on behalf of the LSWR who had to subsequently maintain it. The fact was they did not, for in December 1910 the vicar the Rev. J. Seymour Allen wrote a letter of complaint to the railway company deploring its "very bad state which needed to be remetalled." The LSWR tried to put the onus on to the local highway authority who quickly pointed out where the responsibility rested. By the end of February 1911 this responsibility had been accepted by the LSWR who said that the documentation had escaped their notice. Presumably the damage was repaired soon after. On the closure of the line the anachronism of the railway company maintaining a highway to the station remained. It was resolved in October 1940 when the County Council took over the responsibility for maintenance in return for being given a small parcel of land. The road cut across open fields and crossed the railway by a level crossing at the Basingstoke end of the platform. It was at this point that the only passenger fatality, previously referred to, occurred. Immediately after passing the yard, the road turned at right angles parallel to the railway and rejoined Hackwood Lane. This road had continued straight on and crossed the railway by another level crossing some 340 yards south of the station. Access to the goods yard was from the railway maintained road whilst passengers walked along a narrow pathway to a separate gateway at the rear of the platform at the Basingstoke end of the station buildings.

Opposite the station on the western side of the line an area of land was purchased at a cost of £59.7s for the building of staff accommodation. The housing con-

The wind driven water pump and storage tower at Cliddesden when operational. A similar installation served Bentworth and Lasham.

Builder's plate on the reconditioned water tank.

sisted of a terrace of four brick cottages which fronted on to New Road, whilst set back at right angles to the terrace was a detached dwelling for the so called "Station Agent". The terraced cottages each had an outside coal store and an earth closet whilst the Agent's house boasted an outside w.c. and a coal store. Status was rewarded! The cost of building each terrace house was £258.4.8d, whilst the Agent's superior residence was £426. The cottages were occupied by both station staff and permanent way employees, the occupants in 1911 being Messrs W. Childs, H.W. Broadley, C.G. Knibbs and F. Wright. Rents were £9.15s per annum for Mr. Childs whilst the others paid £7.16s a subsidy of £1.19s being paid by the engineer's department. These figures should be compared with wages - a porter would earn about £1 a week whilst the Butts Junction signalman earnt 22/- per week.

Proper drainage and water supplies were provided from the outset with a 6in. storm pipe leading to a 10 foot deep cess pit 50 yards the Basingstoke side of the level crossing. Water was obtained from a well driven by a wind pump opposite the station and at the bottom of the Agent's garden. The windmill was a feature of the open landscape and was in place until the late 1940's. A brick built tower supported a large tank from whence supplies were taken as required. The tank was supplied second hand (as were all on the line) and renovated; on its side was a cast iron plate "L.S.W.R Wimbledon Works 1900". In June 1929 an agreement was entered into between the railway company and Mr. W.P. Linscott, the owner of a local nursery for the supply of water. The nursery installed a small pump in the 'tank house' and for the use of this space and the supply of water paid an annual sum of £14. Unfortunately this arrangement resulted in shortages to the station and houses during dry periods. Nevertheless the business afforded an amount of traffic for the line and included 'perishables' such as tomatoes and cucumbers which were despatched to destinations far and wide.

To manage the station a Station Master or "Uniformed Agent" was appointed. This was Charles Bushnell who was promoted to the post from that of Inspector at Basingstoke. He had joined the LSWR in September 1868 and was in charge of the station through its first period of operation. On appointment he was paid £70 p.a. together with a rent-free house, free coal and lighting. He finally retired during November 1919 at the age of 71 having served the company for 51 years. In view of his long service he received a gratuity of £60, whilst his pension was 14/6 a week.

A slightly later view of the station from a similar angle, with the usual number of young people beside Mr Bushnell and the 'lad porter'. Above the foliage (which has grown considerably since the previous illustration) at the end of the platform is the loading gauge over the access line to the goods yard. Worthy of note is the concrete platform, a feature of the line long before the LSWR started making its own products at Exmouth Junction. *(Lens of Sutton).*

Devoid of passengers and therefore revenue, Cliddesden has taken on a decidedly forlorn experience in this illustration of the station in its latter years. The gate giving access to the platform has been replaced as has the nameboard. An '02' 0-4-4 heads a train bound for Basingstoke. *(H.C. Casserley).*

Photographs of the period show Mr.Bushnell wearing a frock coat and cutting a very smart figure. That he was held in high esteem cannot be doubted and he took a noticeable pride in the appearance of the station with cornflowers cultivated on the grass bank opposite the platform whilst the name of the station was picked out in chalk. To assist him in his duties a 'lad porter' was also provided. Details of this incumbent are scanty and it is probable that numerous people held the post. By 1912 the post had been re-styled as 'Office Porter' and it was occupied by one Fred Goodall. Duties were many and varied of which one was to attend to the Hackwood Lane level crossing whilst Mr. Bushnell attended to those over New Road.

On 6th April 1916 Leslie Adler became the Office Porter at the rate of 18/- per week. He was born in June 1896 and had joined the LSWR in June 1911 as a signal lad at Guildford. A spell at Effingham Junction followed before his move to Cliddesden. As a young man he must have found the station too quiet for only two months later on 18th June 1916 he resigned as he was "not satisfied with the place". He subsequently rejoined the LSWR in 1920 at Farnham before leaving again in March 1923 to join Oldham Athletic Football Club - quite a move from sleepy Cliddesden.

He was replaced by Alfred Vince on 10th July 1916. Born on 31st August 1896 he joined the railway on 13th June 1910 as a messenger at Basingstoke. Promotions followed and at Cliddesden as Office Porter he was paid 19/- a week. He stayed on after the services on the line had ceased until 5th March 1917 when he went to Wanborough, returning to look after Herriard in 1920 during the 'rail-less' period.

On re-opening, the station was put under the day to day control of a Grade 1 Porter. From October 1924 to November 1925 Cecil Ewing held the post before he was transferred to Bentworth and Lasham. His replacement was William King, the porter who had been reprimanded over the level crossing incident at Bentworth the previous year. His time in the post was short lived for in September 1926 he became a signalman at Basingstoke being replaced by Henry Bennett. He came from a railway family with his father being an Inspector at Basingstoke. He had joined the LSWR in August 1921 at the age of 14 ½ and held various posts at Basingstoke until he was promoted to Cliddesden as a Junior Porter on 12th February 1925 at a wage of 30/- per week. He stayed until 4th March 1926 when he went to Herriard. This was followed by a short period at Basingstoke until he returned to Cliddesden as a Grade 1 Porter on 27th September 1926. Moving on to become a signalman, he was replaced by Bert White on 18th August 1927, the latter being paid at the rate of 46/- per week. The higher wages in the post-war era are a measure of inflation, not an improvement in the financial rewards of railway staff.

Bert White came from Herriard but he had also served at Bentworth and Lasham and was the member

Bert White in charge of Cliddesden in 1932 surrounded by posters exhorting travel to coastal resorts.

of station staff who had the longest service on the light railway in Southern Railway days. He had the distinction of being interviewed by the London 'Evening Standard' at the time of the closure of the line to passengers. This was the second time London reporters came to Cliddesden - the first being the 'Daily Sketch' when the line re-opened in 1924.

In Southern Railway days the additional staff members were classified as Junior Porters. Apart from the aforementioned Henry Bennett other occupiers of the post were George Johnson and Wilfred Bone. George Johnson held a unique record for he was the only native of Cliddesden to work at the station, joining the Southern Railway in October 1925 at the age of 15. The final member of staff was Mrs Violet White who was the daughter of the former Station Master Charles Bushnell. She was 'porter in charge' until May 1936 when she was made redundant.

Traffic at Cliddesden was always small. Indeed with such a small population it was hardly sufficient to support a station at some distance from the village. Passenger traffic was always scanty as so many villagers cycled or walked into Basingstoke - it being so relatively close. Even so, a varied amount of goods traffic was handled which included timber, livestock, agricultural produce and coal. There is no reference to a locally based coal merchant and it is probable a Basingstoke merchant had a consignment sent to the station when he intended to deliver in the area. Bearing in mind the five staff houses it is likely a railway wagon of coal was semi-permanently based in the sidings. There is no record of milk being handled as this would have been delivered by horse drawn cart from the farms to Basingstoke. In later years quantities of roadstone were delivered to the yard in consequence of the

The mixed train of latter years unusually formed with the single passenger coach towards the rear. The usual '02' class tank is seen arriving at Cliddesden with the Basingstoke train. The almost standard feature of locomotives facing Alton will be noticed in this and all the other illustrations

general metalling of the public highways in the area.

The status of the station after the line closed in 1916 is unclear. Originally it was reported to be served by the motor lorry service, but by January 1918 in the 'South Western Railway Magazine' it was reported to be entirely closed. The puzzle is compounded by the fact that Mr. Bushnell did not retire until November 1919, so what did he do and was the station really open? What is certain is that the cottages became vacant and were let to non-railway staff, a situation which had to change when the line re-opened. This was not the only planned change, for in 1923 thought was given to providing a passing loop at the station. This would have gone through the site of the platform whilst a replacement single platform would have been provided at the bottom of Mr. Bushnell's garden. No signals are shown on the plan and it is possible the facilities were intended to be utilised as a goods loop. It is certainly an unusual proposal bearing in mind the line's opening had been forced upon an unwilling Southern Railway.

Passenger figures for the years 1925 and 1927-31 have survived and show a steady decline from an already low starting point.

The high percentage of privilege, or staff tickets reflect the fact that all the cottages were occupied and there was a steady market for the railway's services by the staff, if not by many others. The poor patronage of the line can be seen from the ticket sales of full price tickets which varied between 9 per week in 1925 and just over 3 a week in 1928. The impact of the bus services started by Venture Limited in 1927 can be clearly seen by the poor receipts from that year onwards. It is not surprising for Cliddesden had a remarkable service; six buses every weekday except Wednesdays and Saturdays when there were eighteen and twenty respectively. Four irregularly spaced trains, even offering cheap day returns on market days was not a viable alternative.

Tickets Issued

1925	**1927**	**1928**	**1929**	**1930**	**1931**	
378	233	149	200	193	195	privilege
444	274	174	235	227	229	ordinary
822	507	323	435	420	424	Total

Tickets collected

1925	**1927**	**1928**	**1929**	**1930**	**1931**	
739	508	298	315	344	333	Total

Cliddesden station ten years after closure with nature starting to take over although the platform and loading dock are still quite visible.

Basingstoke Road, Herriard soon after the railway opened. Today this rural lane is the busy A339 and no photographer would dare stand in the middle of the road. *(Rural History Centre)*

CHAPTER 9
THE STATIONS – HERRIARD

Without a doubt the most important of the three intermediate stations on the line, Herriard was placed in typical railway branch line fashion some distance from the village. Access was from Bagmore Lane an important country road linking numerous surrounding villages. In size there was little difference in population to that of Cliddesden, the parish recorded 351 inhabitants in 1901 - a figure which was maintained through the years the line was open. The village population had peaked at the time of agricultural prosperity having 515 people in 1851, but from that period there was a steady out-migration through the latter half of the 19th century.

Dominating the area was the magnificent Herriard Estate the origins of which can be traced back to 1240. When the line opened it totalled some 5,800 acres of prime agricultural land and woodland. Owned by the Jervoise family the Estate was the primary economic force in the village employing up to one hundred staff at its zenith. As with Cliddesden, Herriard village was for many years an almost self contained community with employment centred on two major areas - agriculture and estate work. With numerous tenant farms, agricultural practice varied with both arable and pasture leading to considerable quantities of milk being produced which was despatched in churns from the local station. A considerable domestic staff was employed in 'the House' whilst the Estate also had its own sawmill which was powered by a stationary steam engine. Consequently there was a steady supply of coal for the sawmill whilst cut timber and timber products were despatched. The Estate generated considerable traffic with seeds and fertilisers coming in and corn, hay, straw and livestock being sent. The 'House' saw everything delivered by rail - from manure to oysters!

The first coal merchant to use the railway yard was Messrs Stephen Phillips and Co. of Basingstoke with deliveries being made in the local area from stock within the yard. However by June 1911 further ground was made available to R. Toomer and Co. for a coal depot. Due to the limited trade in the surrounding area it is not surprising that neither of these firms had railway wagons branded as being based at the station.

Situated mid-way along the branch Herriard was unique amongst the stations on the line in as much as it originally boasted a passing loop served by two platforms. As a tablet station trains could pass as required. Protection came from two home signals, one at each end of the station, which afforded access into the station for trains from either direction. The 'down' signal (from Alton) was within the station limits but the 'up' signal was on the Basingstoke side of the level crossing with Bagmore Lane thus protecting it. This crossing, like all the others on the line was hand operated, the gates being hung on stout concrete posts. Access to the station was from Bagmore Lane by two separate gates. That closest to the crossing was used by passengers who would walk along a path parallel to the rear of the platform - similar to the arrangement at Cliddesden. Access to the platform was through a double gateway adjacent to the station building.

The station building was on the 'up' side and served trains from Basingstoke when crossing was necessary. An identical structure to Cliddesden was provided being built of corrugated iron with a match boarded interior. The platform surface was similar

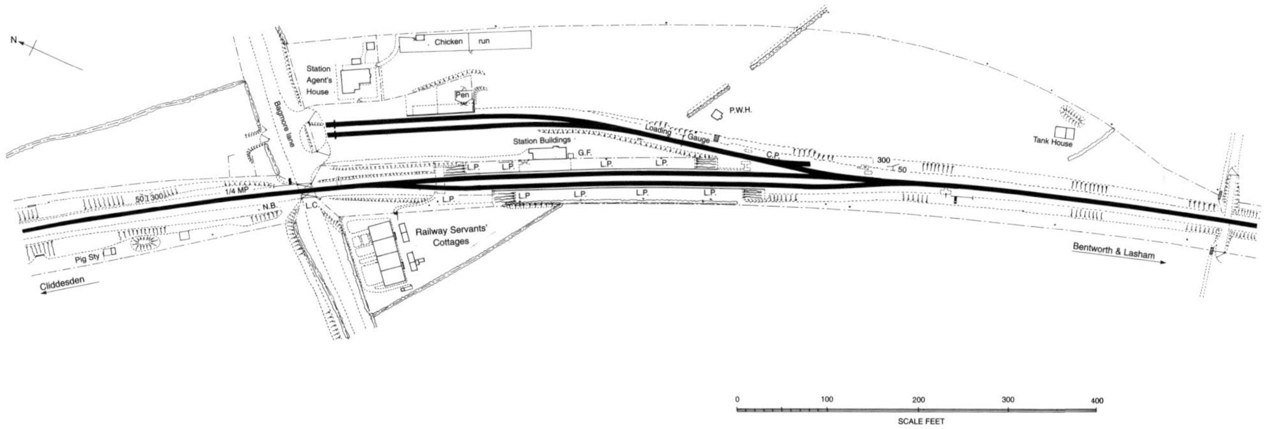

HERRIARD - 1911

Herriard station looking towards Alton in the early years with the original name board in situ. Mr. Hooper the stationmaster is in evidence but traffic is limited to a few milk churns.

and was lit by four standard LSWR 'candy twist' lamps with a fifth lamp over the station entrance. Opposite, although slightly staggered towards Alton was the second platform, access between the two being by a board and cinder crossing at the Basingstoke end of the station. Illumination came from a specially provided lamp. No shelter of any kind was provided on the 'down' platform although it was lit by four lamps. The Parish Council had appealed to the LSWR in 1901 to provide a shelter, but the request was refused. The platforms were of concrete construction as they were at all the stations on the line. A material which later became closely associated with the Southern Railway, it must have been one of its first uses on the LSWR.

During the first part of the station's life the tablet instrument was probably located within the booking office as no signal cabin was provided. Train movements were controlled by an open air lever frame on the 'up' platform adjacent to the Alton end of the station building. Although no official information has been found as to the interlocking arrangements, it would have been necessary for the release of one lever to lock the other so as to prevent oncoming trains using the same loop. It is believed that the platform lever frame also operated the turnout leading into the yard. The yard itself was identical to Cliddesden with two sidings, one being served by a loading bank and livestock pen. The turnout within the yard was operated at ground level and a loading gauge was located at the exit.

Living accommodation for the staff was of identical design to the other stations. The detached Agent's house was adjacent to the goods yard and from it paths led to a shed and a substantial chicken run; the latter being parallel to the railway company's boundary. Four terraced cottages stood on the opposite side of the line and fronted on to Bagmore Lane. They were linked to the station by a separate pedestrian access which joined the crossing linking the two platforms. All of these properties were built of mellow red brick under a slate roof and boasted the modern amenity of an indoor sink. Outbuildings were the same as at Cliddesden although a pig sty was built adjacent to the line on the top of the cutting to the north of the station. The cottages cost £271.6.4d. each to build (slightly more than Cliddesden) and the price of the land was the princely sum of £38 for all four. The Agent's house cost £426 (as at all the other stations) whilst the land was purchased for just £19. At the south end of the station, close by the line leading to the yard was a permanent way hut and beyond this, high above the running line was a water tank house; supply problems meant that a well 300 ft. deep costing

A similar view taken slightly later when the vegetation behind the fence has grown and the station nameboard has been changed. Under a glass it is just possible to discern the open air lever frame on the platform although detailed study is rendered impossible due to the proximity of the station furniture. A number of milk churns are apparent as are what appear to be fruit boxes. Obvious by their omission are any passengers. Beyond the right hand platform it is just possible to see the rear of the protecting signal on the opposite side of the line to the water tower. *(L.G.R.P.)*

An interesting comparison taken in the final years of the line with just one platform in use. In the background at least two coal wagons stand in the yard whilst in the doorway to the station building are the 'porter in charge' and a child. Compared to the previous photograph the trellis fence has been replaced by Southern Railway concrete post and wire and it is now possible for the public to use the telephone at the station. Beyond a row of fire buckets have appeared, a cast iron warning notice warns passengers not to trespass and the water tank now has a protective roof. Noticeable is the lean-to gents lavatory and the inevitable group of milk churns. *(H.C. Casserley)*

Station Master Herbert (Harry) Brazier and driver Sidney Tett loading the Clayton motor lorry in 1918 with milk and agricultural produce. The chain drive is clearly visible. On the platform is a sack barrow with the station name just discernible and behind is the garage specially built to house the lorry.

Schoolchildren at Herriard c.1930. The wooden shed used by the road motor service is visible to the right as are the fire buckets mentioned earlier. In the background is Bagmore Lane level crossing which was under the control of the man in charge of the station. Just beyond the open gate on the platform a solitary coal truck occupies the yard.

A family snapshot; local residents Mr Hart and Charlie Carter sitting in a coal wagon at Herriard station.

£700 had to be sunk which was operated by a pump powered by an oil engine. The well was the scene of an unfortunate accident in 1908 when the regular fitter, being absent, his assistant a Mr. Butler volunteered to go down and carry out maintenance work. As he was being lowered over the side a false move unhooked the seat causing him to fall to the bottom with fatal consequences.

Herriard, unlike its neighbours, was a grade 6 station - the other two being 'unclassified'. In charge as first uniformed Agent was Arthur Hooper who was promoted to the post from that of platform inspector at Woking. His salary was identical to his other colleagues on the line - £70 p.a. plus rent free housing and coal. In June 1910 Mr. Hooper retired at the age of 69 and was replaced by W. Hazzard a goods guard from Yeovil Junction. He had joined the LSWR in May 1872 and his presence continued the practice of putting the stations under the control of experienced staff. Unfortunately his period of office was short for he died in office on 9th February 1915. In view of his long service his widow was awarded a grant of £20 from the company. The replacement in April 1915 was Albert Russell, appointed at a wage of 27/6d a week, a rise of £1.5/- p.a. over his predecessor! He stayed until March 1917 when he was promoted to Elsted, a class 5 station on the Midhurst branch. As such his salary rose to 30/- per week.

In this early period it is probable that the Agent was responsible for the operation of the tablet instrument and platform lever frame. Meanwhile the staff would have operated the crossing gates over Bagmore Lane.

In 1917 when the service over the line had ceased the LSWR sought to consolidate its staffing position and it transferred the Station Master from neighbouring Bentworth and Lasham which was to close entirely. Consequently Herbert Brazier took charge at a wage of 27/6d a week during the period when the road motor service was in operation. He had joined the LSWR in September 1882, and again, was a well established employee. War time inflation was reflected in Mr. Brazier's salary for by August 1919 he was earning £160 p.a. plus a £34 bonus. He was to be the last Station Master on the line for in November 1919 he was transferred to Ash Green on the Guildford to Farnham direct line.

Details of the other staff employed at the station during the LSWR period are incomplete and it is probable there was a regular turnover of staff. Certainly details of permanent way staff have not been located. Between 1910 and 1913 Arthur Hunt was the Office Porter and Relief Signalman and he was joined in early 1912 by Charles Nelson Bushnell, son of the Cliddesden Station Master. He had been born in November 1890 and as a boy must have seen the early trains on the light railway. No doubt influenced by his environment he joined the LSWR when he was not quite 14 moving around the system in Hampshire and Surrey. He left Herriard in August 1913 to become a signalman at Basingstoke.

The station's longest serving member of platform staff in the pre-grouping years was Sidney Bunnett who served between March 1913 and June 1916. He was no stranger to the line having been at Bentworth and Lasham between 1905 and 1908. His pay at 22/- per week was better than that in agriculture and was considered reasonable for the period.

Maurice Cockwell made a brief appearance in 1915 and he was replaced by Harry Knight who lasted until 8th February 1917. Another member of staff in 1916 was W. Ricketts who resigned in 1917 just after the service was withdrawn probably fearing that the job prospects were poor in the extreme.

From 1917 the traditional post of porter was not filled, it being replaced by that of 'Motor Vehicle Driver' who took charge of the railway operated replacement lorry service. The LSWR with few vehicles and drivers therefore not only transferred the vehicle from Devon, it transferred its driver as well. Sidney Tett was born in December 1887 and joined the LSWR as a junior porter at Exeter. By September 1911 he was the ticket collector on the Chagford bus service, being promoted to driver in February 1913 at the wage of 28/- per week. At the end of 1916 his wage had increased to 35/- a week - a not inconsiderable sum and more than the Herriard Station Master! With the cessation of the bus service he and his vehicle moved to Herriard in February 1917. There he stayed until May 1920 when he returned to Devon to take up his old job as the Chagford bus driver. His wages reflect war time inflation for when he returned to Exeter he was earning 79/- per week.

His replacement was William Dimmick who had joined the LSWR at Basingstoke in 1919 as a porter. Promotion followed quickly for he was elevated to a Grade 2 Porter on 1st January 1920 and then to 'Motor, parcels and van man' on 27th May when he went to Herriard. He remained there until the motor service was taken over by Mr. Gilbert when he and his vehicle went to Crewkerne on 1st January 1923. It seems that the LSWR had a policy of 'have van must travel' and one can only sympathise with Mr. Dimmick

'02' class No. 234, entering the station with a Basingstoke train on 13th June 1931. Despite the lack of traffic, or perhaps because of it, the site is neat and tidy with no obvious litter and weeds in the space previously occupied by the second line of rails. *(H.C. Casserley).*

The daily goods at Herriard in 1927 with a somewhat light load. Posing for the camera are junior porter Bert White, driver George Appleford, Grade 1 porter George Stickland and guard Walter Young.

The railway cottages at Herriard in 1986. Somewhat surprisingly the windows and doors have not been altered and the terrace is as built apart from the two single storey extensions at each end. Beyond is the Station Master's house which emphasises the difference in status associated with the positions on the railway.

on his long journey down to Somerset with a solid tyred lorry.

Following the transfer away of Herbert Brazier the last Station Master, the LSWR still needed a responsible individual at Herriard to look after the day to day business. After a hiatus of several months when the job was probably filled by relief staff from Basingstoke, Alfred Vince was appointed Office Porter on 4th March 1920. It will be recalled that Vince had been at Cliddesden on the closure of the line and had been transferred to Wanborough. Now he was called back to look after the LSWR's station without a railway. He, like the driver William Dimmick stayed until Gilbert took over the cartage and having safely transferred the operation was moved on, initially to Basingstoke and then to Privett on the Meon Valley line. He retired from his railway career in 1961 as Station Master at Romsey.

The only physical change to the station after 1916 was the construction of a garage for the vehicle undertaking the delivery service. The building was probably alongside the rear of the Alton end of the 'up' platform. At the same time it is believed a store shed was also provided. The reason for this structure is unclear as accommodation would have been available within the station buildings which were of course effectively out of use. The only possible use for

Edith Oliver and her daughter Dorothy at the door of their railway cottage.

Devoid of trains for over a decade the station site is still tidy. To the right the garage is still standing amongst the trees. The staggered nature of the platforms is apparent in this 1947 view looking towards Basingstoke.

this additional structure was for the storage of fuel, oil and vehicle spares.

It has been seen that the losses associated with the running of the railway replacement service led the LSWR to plan an alternative strategy and as a result they agreed to privatise the operation. Although the decision was taken in May 1922, somewhat surprisingly it was not implemented until 1st January 1923. It could well be that the LSWR Board, recognising that their days were limited decided to give the problem of the road operation and the official closure of the rail service to the newly formed Southern Railway. In this they made a wise decision.

As recounted the newly formed Southern Railway were compelled to re-lay and re-open the line in 1924. The differences of opinion between the traffic and engineering departments have been mentioned previously and the surviving correspondence points to a farcical and costly situation at Herriard. It would appear that when re-laying was in progress the passing loop was relaid but not connected. Evidence of this extraordinary situation comes from the letters of the time. When details of the train service became known and it was realised that the loop was superfluous, it was lifted soon afterwards.

With no passing loop Herriard was no longer a tablet section and consequently the arrangement for entering the goods yard was altered. The single set of points affording access were controlled by a lever ground frame alongside the running line. This was released from the tablet which now covered the whole line from Basingstoke to Butts Junction. Other changes took place in the goods yard. It would appear that Mr. Gilbert's access to the shed was terminated in 1924 as he could only reach it over the newly relaid sidings. There is no clear reference as to whether this termination referred to just the access to the shed or to Gilbert's tenancy and business with the railway at Herriard. Probably it was the latter, for with the railway restored the Southern Railway would not want to loose the smallest amount of traffic to a competitor, let alone to one it had installed itself.

A new staff had to be found for the re-instated service and the Southern Railway allocated a Grade 1 Porter and a Junior Porter to the station. George Stickland was the former coming to Herriard on 19th February 1925. He had joined the LSWR in 1917 at the age of 14 and had a variety of jobs at Winchester, Eastleigh and Basingstoke before being appointed to the branch. It is probable that he came to Herriard from Basingstoke on a 'relief' basis, the formal appointment being made several months later.

One of the first Junior Porters was Edmund Leadbetter who, unusually was appointed straight to Herriard.

Usually the railway company only appointed staff who had some experience as they had to be able to turn their hands to most things and cope with a variety

Forty years later nature has taken over and it is only just possible to identify the platforms under the ground cover. A view from a similar position to that previously but in 1986.

of situations. He lasted from September 1925 to February 1926 when he was replaced by Henry Bennett from Cliddesden who gave the station the experienced pair of hands it had lacked. The next eighteen months saw a regular turnover of staff. Henry Bennett went off to Basingstoke before returning to take charge at Cliddesden, as previously recorded He was replaced by David Lewis on 28th June 1926, a native of Basingstoke who joined the railway at Herriard on his 16th birthday. Seven months later he was transferred to Bentworth and Lasham whilst Bert White came from there in his place. He, it will be recalled, went on to take charge of Cliddesden in August 1927 and was replaced by George Johnson, previously at Cliddesden. Wages for the mid 1920's had stabilised after the War and a Grade 1 Porter could expect to earn 50/- per week whilst an experienced Junior Porter received up to 35/-. Newcomers like David Lewis however only received 20/- although they could expect annual increments of 5/- a week to bring their wages up to the maximum after three years' service.

Initially passenger traffic levels were good, especially when compared with the low population of the parish and it was quite obvious that Herriard was used as a rail-head for the numerous villages surrounding it. One anomaly as at Bentworth and Lasham, was that the number of tickets collected always exceeded those issued, whereas at Cliddesden the situation was reversed. The question as to what happened to all those one way travellers remains a mystery!

Tickets Issued

1925	**1927**	**1928**	**1929**	**1930**	**1931**
3517	2839	1167	816	823	688

Tickets Collected

1925	**1927**	**1928**	**1929**	**1930**	**1931**
3550	3315	1483	1026	1411	784

Again it will be seen that the impact of the competing bus service from 1927 was immediate and led to a steady terminal decline. The inconveniently sited station with an erratic service on weekdays only could not compete with Venture's service which went through the centre of the straggling village. From the beginning a late bus operated on Saturday evenings allowing the villagers a taste of Basingstoke night life, up to 10.40 p.m. at least! The declining receipts were

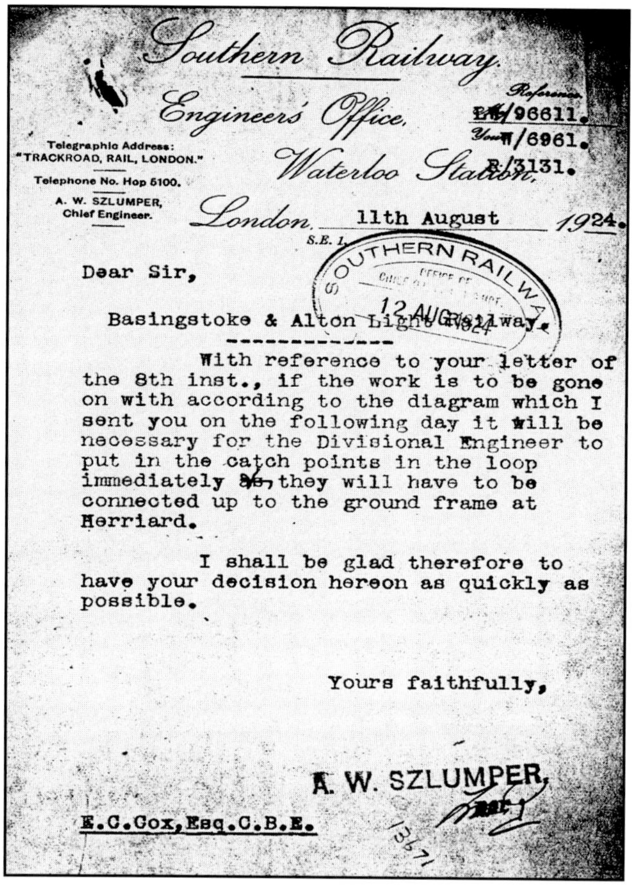

Southern Railway.

Engineers' Office, Waterloo Station, London, S.E. 1.

Telegraphic Address: "TRACKROAD, RAIL, LONDON."

Telephone No. Hop 5100.

A. W. SZLUMPER, Chief Engineer.

Reference LW/96611.
Your W/6961.
R/3131.

11th August 1924.

SOUTHERN RAILWAY OFFICE OF ... 12 AUG 1924

Dear Sir,

Basingstoke & Alton Light Railway.

With reference to your letter of the 8th inst., if the work is to be gone on with according to the diagram which I sent you on the following day it will be necessary for the Divisional Engineer to put in the catch points in the loop immediately ~~as~~, they will have to be connected up to the ground frame at Herriard.

I shall be glad therefore to have your decision hereon as quickly as possible.

Yours faithfully,

A. W. SZLUMPER,

E.C.Cox,Esq.C.B.E.

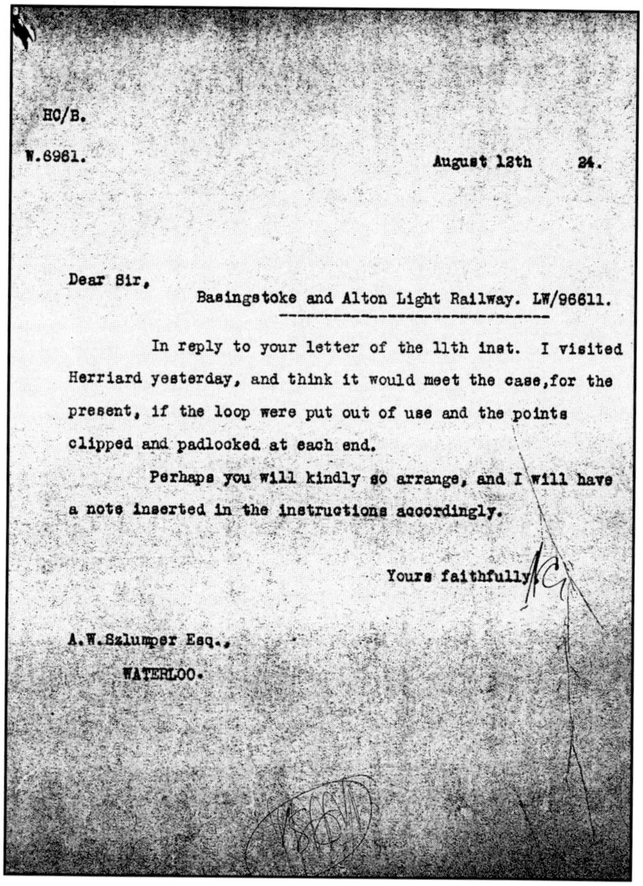

HC/B.

W.6961.

August 12th 24.

Dear Sir,

Basingstoke and Alton Light Railway. LW/96611.

In reply to your letter of the 11th inst. I visited Herriard yesterday, and think it would meet the case, for the present, if the loop were put out of use and the points clipped and padlocked at each end.

Perhaps you will kindly so arrange, and I will have a note inserted in the instructions accordingly.

Yours faithfully,

A.W.Szlumper Esq.,
WATERLOO.

The Herriard loop relaid only to be taken out of use immediately.

minimal with only 13 tickets a week issued in 1931 which equates to approximately one sale for every two trains calling. Of these just under 13% issued were at 'privilege' or staff rates.

With the sudden decline in passenger activity in 1931, closure to passengers in 1932 could hardly have been a surprise. Milk traffic was always important even in these later years with the 1929 working timetable specifically relating to this and the working of empty churns back from Forest Hill, London.

The final action of the Southern Railway was to connect the houses to the mains water supply so they could decommission the well, pump and tank house. This operation cost £75 but the company recovered materials to the value of £20 and overcame the long-term maintenance costs. Following the work, the railway sold the houses to private owners. An indication of current prices is the fact that one of the cottages was sold in 1997 for a price in the region of £120,000.

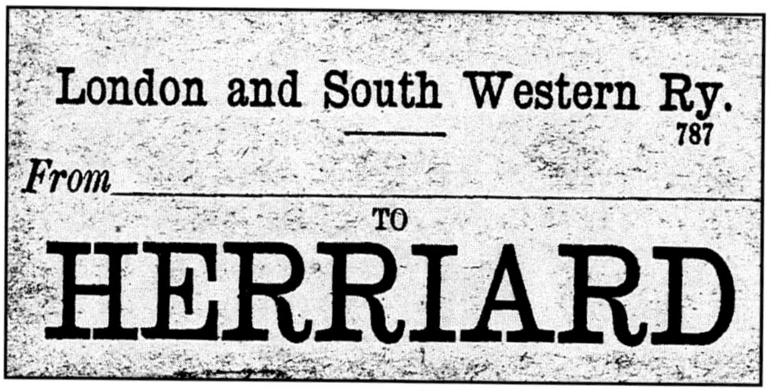

London and South Western Ry.

787

From ____________

TO

HERRIARD

CHAPTER 10
THE STATIONS – BENTWORTH AND LASHAM

The third station on the line and the closest to Alton was Bentworth and Lasham. Situated in a shallow valley on an exposed hillside overlooking a minor road it was not well sited for local traffic. The village of Lasham was nearer being only ½ mile away. However it was situated on a hilltop and was 100 feet higher than its station - fine if one was going to catch a train but not so good for the return! Its parish population in 1901 was only 144, the village having declined in size throughout the Victorian years from a high point of 284 in 1841. Bentworth, in comparison was larger but further away and the climb out of the valley was more severe. When the line opened its parish population was 571 - a figure which was significantly lower than its peak of 647 in 1861. The village was the largest in the area but its relative decline in the 19th century can be seen by comparing its population with that of Alton. In 1801 Bentworth was 20% the size of Alton; in 1901 it was 10%. The comparison with Basingstoke is even more startling. In 1801 its population was 16% that of its northern market town, by 1901 it was only 6%. The move to town from village which took place throughout this period in Britain was quite clearly taking place in northern Hampshire. Although both villages had their share of 'service industries' such as wheelwright, post office and carpenter the predominance of agriculture as an employer had led to stagnation and relative decline especially when compared with its market town neighbours.

Railway facilities were similar to those at Cliddesden with two sidings behind a single platform. The only difference was that they faced Basingstoke whereas at the other two stations they faced Alton. Again a loading dock and livestock pens were provided and a loading gauge was sited at the neck of the sidings.

At the Alton end of the platform was the customary hand operated gated level crossing, in this case taking the main road from Basingstoke to Alton. This road was down graded to the status of a lane when the main road was re-routed parallel to (and eventually on) the trackbed of the line in 1943 when the new Lasham aerodrome severed the original road. Intending passengers gained access to the station through a gateway adjacent to the level crossing via a path which ran behind the rear of the platform to a pair of gates which opened on to the platform next to the station building. This appears to have been identical to those elsewhere on the line and was probably similarly divided within. The platform was illuminated by five LSWR 'candy twist' lamp posts with a sixth next to the gate by the level crossing. A simple name board indicated the location and the platform was bounded by the same type of trellis fence found at the other stations on the line. A feature different from the other locations was the fact that the line curved at this point and the platform had to compensate for this. As elsewhere it was of concrete construction backfilled and surfaced as before. The station site was unusually large with the boundary allowing for considerable expansion of facilities if required. Very little of it was.

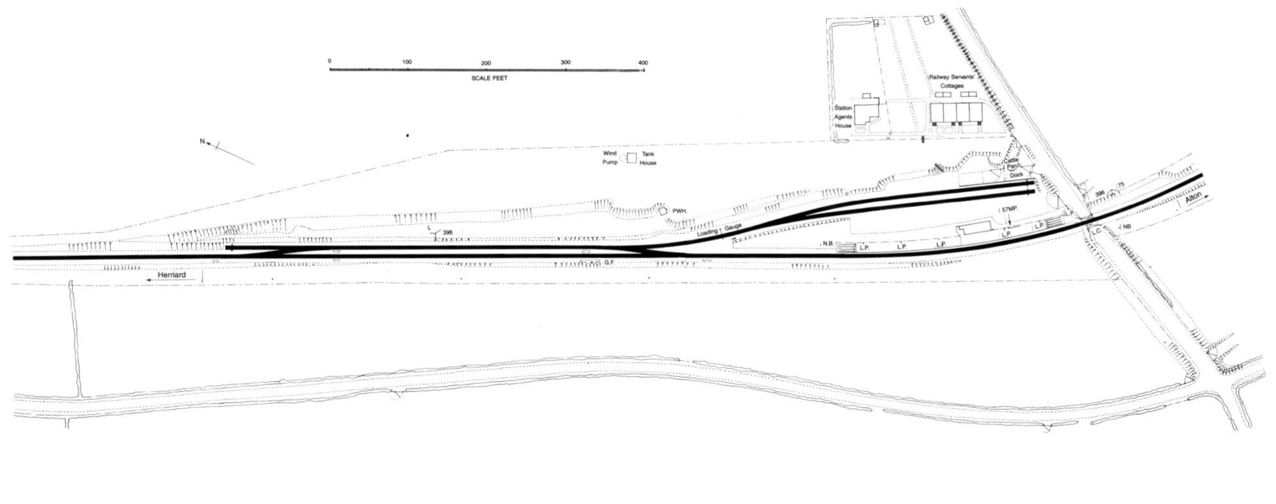

BENTWORTH AND LASHAM - 1911

This was the only station to alter during the first period of operation, with a new loop siding being installed at the Herriard end in March 1904. At the same time the access to the sidings was altered to feed into the loop rather than the running line which had been the case. In theory a train could have been 'shut in' to shunt the yard whilst another passed on the single

Probably taken just after the line opened this view of Bentworth and Lasham gives a good impression of the isolation of the station. Compared with the next view no trees can be seen although what is just visible is the wind driven pump.

(Lens of Sutton).

A slightly later view of the station, its staff and their housing showing the station as built with trellis fence and original nameboard. The road linked the two villages which gave the station its name. *(Lens of Sutton).*

running line. However as this was not a tablet station only one train could be at the station at a time, so this facility could not have been used in normal operating circumstances. An open air ground frame of four levers controlled the site and this was released by the tablet. No fixed signals were provided. The reasons for this increase in facilities was reported as necessary so as "to facilitate the working of goods traffic". In reality this allowed Alton facing locomotives to use the loop to run round their train and then shunt the sidings. The cost of this luxury (the train could have shunted the yard on its return from Alton) was £592.

The new siding layout and loop were inspected by Major Pringle of the Board of Trade on 28th April 1904 and passed as fit, his only comment being "in view of the fact it is a light railway I do not consider a bar is necessary". This reference was to the lack of a fouling bar to prevent the points being moved when a train was standing on or near them; this being at the time a standard feature of signalling installations. A surprising feature of pre-1904 plans is a sketch of a "road goods box" sited on the platform at the Basingstoke end of the station building. The existence of this structure must be open to question bearing in mind the date and the limited traffic. This was clearly demonstrated by the fact that the station closed entirely at the end of 1916.

Facilities for staff accommodation comprised the customary Agent's detached house and a row of four terraced cottages - all identical to those at the other stations. These cost the same as at Herriard, namely £271.6s 4d. per cottage and £426 for the house. All stood next to each other, parallel with the running line and the yard. The plot, which cost £144.10s was in an elevated position and consequently the housing overlooked the station site. Staff access was by a path with two sets of steps which led into the yard or alternatively by the main yard access which was between the houses and the level crossing. Being somewhat exposed trees were planted around the housing boundaries on 2½ sides. Water was drawn from a well operated by a wind pump as at Cliddesden. This was stored in a reconditioned tank atop the customary brick tank house located alongside the boundary fence high above the loop siding on the side of the line. Below it, but still above rail level was a

A view seen across the fields from what is now the A339 but what was then a country lane. The photograph shows the facilities to advantage and was taken at a similar time to the previous picture. The station building and the staff housing at a higher level are clearly identifiable with the distinct separation between the terraced cottages and the Station Master's house being particularly noticeable. To the right is the main road giving access to the station just beyond the level crossing, the pedestrian and yard access being guarded by a solitary lamp. Within the yard are at least four vehicles, two open trucks, a closed van and a cattle wagon. The conifers which have just been planted in front of the station as a windbreak would eventually become highly effective.

From the level crossing looking towards Basingstoke during the second period of operations. The nameboard has been changed and the trees have grown somewhat!

permanent way hut. Besides providing domestic and station supplies, water was also stored in 10 gallon churns to be used to replenish the tank of the train engines if required. This facility was used only occasionally, mainly in the last years of the goods service when the station was the terminus of the line and the original practice of refilling at Alton was no longer available.

In charge as the first station Agent was Mr. F. Spain (not Pain as recorded by the local newspaper) who had been promoted from Inspector at Southampton West. He, like his colleagues, earned £70 p.a. and had the use of the house with lighting and coal rent free. It is likely he was an experienced member of staff like the others, but his movements/retirement remain unknown. By May 1907 the post was filled by William Vicary who had joined the LSWR in July 1867. He retired on 2nd July 1914 to be replaced by Herbert Brazier. He was to be the last stationmaster and as reported earlier he was transferred to Herriard in 1917 when all services ceased. From this period Albert Kneller the Basingstoke stationmaster is referred to as being nominally in charge.

As at the other stations the stationmaster was assisted by a uniformed porter. Information is scarce - the cottages were occupied by Messrs W.J. Oliver, Walter Oliver, H.R. Budd and P. Willis - the latter probably being the station staff as all the others were employed by the Engineers Department. The Oliver family seemed to have dominated the local railway labour market for it is known that Harry Oliver had worked on the construction of the line and subsequently served at Herriard on the permanent way staff. The only known staff member in the LSWR years was Sidney Bunnett, 'Office Porter' from 28th September 1905 to 26th November 1908, being paid 17/- a week. He joined the railway in July 1900 aged $13^3/_4$ and travelled extensively starting in London at Wandsworth Road before coming to Itchen Abbas on the Mid Hants line in 1904. On leaving Bentworth he went on to Addlestone, Farnham and Wimbledon before returning to the line at Herriard in March 1913. The policy of moving staff between the stations on the line, even if they had moved away, was established from an early date.

In the immediate period after re-opening William King was the Grade 1 Porter and it was he who failed to open the gates on 20th November 1924 with the result that the 2.05 p.m. mixed train from Alton hit them. King had been on the railway for seven years and had worked at as many stations - Bentworth and Lasham was his eighth appointment. He, too, had travelled widely filling posts in Hampshire and Surrey. He was transferred to Cliddesden in November 1925 and was replaced by Cecil Ewing, who came from there - the reason for the swapping of staff is unrecorded. Ewing came from a railway family and his father was a Station Master on the Meon Valley line. He was an experienced man who had held posts in nine Hampshire stations and undertaken military service between 1916 and 1919 before coming to Bentworth and Lasham. By 1927 he was living in the Lasham crossing cottage and his wife Gladys was the crossing keeper.

Just as at the other stations a Junior Porter was appointed to help. From September 1925 to January 1927 Bert White was in post before moving on to Herriard and then to Cliddesden. Bert joined the Southern Railway in May 1923 as a 15 year old messenger at Basingstoke. He spent all his working life on the railway and completed nearly 50 years service. During his time at Bentworth and Lasham his pay increased annually by 5/- a week as his experience grew, until when he left for Herriard he was earning 35/- a week.

His replacement came from Herriard and was the young David Lewis. He was unlike his predecessor and his period in office was short, his services being terminated in June 1927 for "irregularities in connection with the issue of blank card tickets". History does not recall his replacement although it is known that the final railwayman associated with the station was porter Symonds who was there in 1936 at the end of the goods only period.

Apart from porter Ewing's wife the only known gatekeeper at Lasham crossing was George Sainsbury. He had travelled around Hampshire and Surrey until coming to the line in April 1906. The closure of the branch meant that he was transferred to Alton in March 1917 from whence he retired on 25th July 1921 with a pension of 7/- a week. Unfortunately he died shortly afterwards in January 1923.

Traffic handled at the station was similar to that at Herriard with agricultural produce predominating, milk being especially important. Coal would have been delivered on a regular basis, but other inward bound traffic would have been associated with farming, such as feedstuffs, seed and fertiliser.

The effects of the heavy snow storms of late December 1927 are shown here to advantage in this view of Station Road, Bentworth. It is just possible to discern the railway housing to the left of the hedge - the station itself looks quite buried! Not surprisingly the railway was impassable for several days and when it was opened it was the only route open for sometime, the roads being blocked. The view shows the climb necessary from the station to reach the village which is some way behind the photographer. *(Hampshire County Museum Service)*

Bert White and friend astride their motorcycle in 1926. It is just possible to see the concrete platform edge through the fencing to the right of the level crossing. *(E. Griffith collection)*

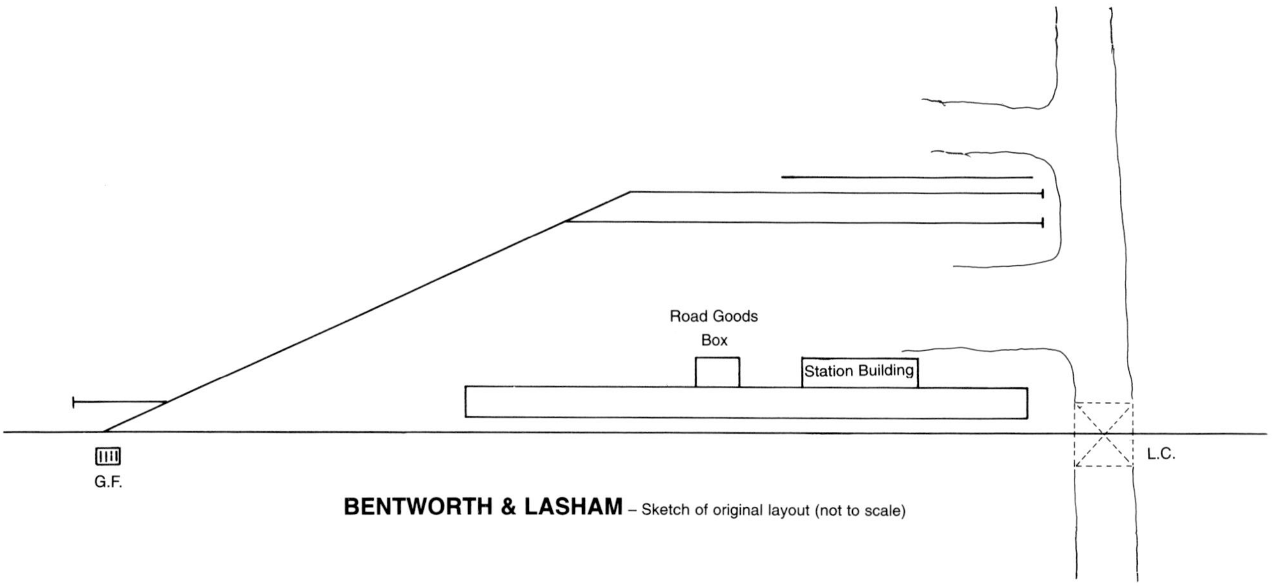

BENTWORTH & LASHAM – Sketch of original layout (not to scale)

A lone figure waits forlornly for the train that will never arrive. This 1947 view shows the station, the level crossing gates and the station fencing all intact and the windbreak still growing well!

Tickets Issued

1925	1927	1928	1929	1930	1931
3117	2251	1162	904	801	745

Tickets Collected

1925	1927	1928	1929	1930	1931
3165	2541	1320	1074	857	846

A glance at the passenger traffic receipts for the final years gives a similar pattern to that experienced at the other stations. After a promising start, traffic fell away in the now established pattern although the fact that it did not decline as fast as Herriard is because the Venture bus service went via Medstead. It took 36 minutes from Bentworth and Lasham station to Alton by bus whereas the train took only 15 minutes. This was fine if one lived near the station or along the valley; otherwise there was that hill to climb! Compared with the other two stations, Bentworth and Lasham recorded the lowest number of privilege tickets issued on the line, being just under 9% of the total. This was due to the limited number of staff employed in the latter period with some of the cottages being occupied by non-railway staff.

Certainly like the other locations it was quiet and this meant there was little urgency to undertake paid duties. In later years with the station "supervised" from Basingstoke, it is probable a somewhat casual atmosphere prevailed. The reality was that the cost conscious and dynamic Southern Railway was not prepared to support such charming rural corners indefinitely.

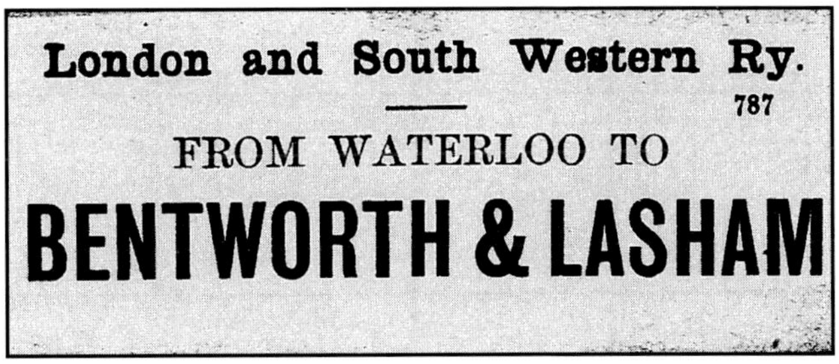

A new use for an old station. The platform makes a convenient loading bank and the trackbed provides an access road. The conifers have gone but the horse chestnuts remain in 1986.

In 1986 the station may be open but not for railway business; the building still fulfils a useful function but is now a shadow of its former self. The cottages behind are still standing but have been extended.

CHAPTER 11
EPITAPH

The Basingstoke and Alton Light Railway has been described as a hopelessly uneconomic branch line whose most profitable activity was its use by film companies. This dismissive statement, which cannot be denied, is superficial. It ignores the reasons behind the line's construction and highly unusual subsequent history as well as the economic climate of the period. The comment also makes no allowance for the technological changes that impacted on the transport industry after 1918.

In many ways the problems that beset the line started at the outset. The LSWR thought it could build the line for £67,000; the reality was that it cost nearly £108,000. This shows the deficiency of the engineering estimate for which Galbraith must take responsibility, but land requisition cost in excess of £22,000 whereas only £10,000 had been allowed. Landowners might have seen the benefit of having a local railway line but they were determined to get the maximum for their holdings. Scotter's wish that land would be sold at market value was nothing short of naive; indeed it turned out to be wishful thinking.

Viables crossing in 1947 with the Basingstoke by-pass crossing the trackbed. It was here that the local farmer was killed in 1911 when the line was crossed, not by a highway but by a farm track on the level. Basingstoke is a long way off in this rural view.

Viables roundabout in 1986. The 1930's by-pass has been superseded by the M.3 and the concrete bridge demolished to make way for a roundabout in urban Basingstoke. A short section of track depicts the line of the railway, all traces having been obliterated.

The argument that the branch was a "blocking line" built solely to stop the GWR penetrating further into Hampshire cannot be sustained. It has been shown that pressure for new railway construction came from Portsmouth and that Paddington showed no willingness to become involved in a territorial dispute with the LSWR. Nevertheless Scotter's willingness to promote the Basingstoke and Alton and the Meon Valley lines immediately after the failure of the third railway scheme in the area cannot be dismissed as coincidence and it is likely that he realised another proposal would follow eventually. Consequently he believed it would be better for the LSWR to take the initiative rather than to be forced into a reactive, defensive position. The fact that the Light Railways Act would allow the Company to build a line "on the cheap" was a timely coincidence which presented an opportunity that could not be ignored. Certainly the magnitude of error in the estimate and the speed with which it was presented to the Light Railway Commissioners shows a certain amount of hurried thinking, but at minimum cost it dismissed the opposition - past, present and future. This can be interpreted as building a political "blocking line", but in reality should be seen as a commercial decision in line with the LSWR's overall policy of paying greater attention to its Hampshire traffic of which it had a dominant position. Without a doubt the decision was based on false data and excessive optimism, but hindsight makes wise men of us all.

It is important to see the branch line in the context of what was happening to the railway industry generally. On 13th May 1910 the LSWR and GWR agreed to pool receipts from competitive traffic thus ending any commercial rivalry that may have existed. Although the railway industry was prosperous in these Edwardian years with increasing net receipts, the Basingstoke and Alton showed increasing losses. The Great War produced contradictory conclusions too. Whilst other lines had reduced services and a small number were indeed closed for the duration of the hostilities, few were dismantled. Of those that were, only the Basingstoke and Alton was re-instated. Consequently the line's early history can be seen to be strangely at odds with what was happening nationally.

The highly effective campaign to re-open the line and the Lord's subsequent findings indicate that the changing nature of the transport industry had not been recognised by those in authority locally, or more importantly, by Government. The attitude that railway companies had responsibilities to society rather than to the shareholders was outdated now that competition was to come from outside the railway industry and not from within it. The words of the Minister of Transport should have been heeded. In 1921 the Rt. Hon. Eric Geddes stated that "the poor financial results obtained from light railways in Great Britain have been mainly due not to the failure to exercise economy in working

Lasham level crossing and crossing keeper's cottage seen from the trackbed looking towards Alton. *(A.E. Bennett)*

Bridge 7, one of the few remaining bridges on the line in 1986. Road improvements have now led to its demolition.

The trackbed still being used to move agricultural produce. A 1947 view with the new main road skirting Lasham Hill to the right of the picture.

expenses, but to the smallness of the earnings in relation to the capital involved". He went on to say that "it must be frankly recognised that a road motor service may be a desirable alternative to a light railway - future policy should aim at co-operation and co-ordination rather than competition". The tragedy was that his words, which were supported by Col. Holman F. Stephens owner of numerous light railways in Britain, were ignored when the Basingstoke and Alton was being debated two years later.

It has been seen that the introduction of a competing bus service had an immediate impact on passenger figures. This was not predicted by the Lords in 1923 although Walker was fully aware of the threat. The fact that the majority of traffic, both passenger and goods was of short distance made it particularly vulnerable to competition. Agricultural traffic, the mainstay of the line's freight business changed radically during the inter war years as the distance of farms from stations meant that rail traffic became less desirable. This was especially so as the motor lorry could provide a door to door service. When the Basingstoke to Alton re-opened in 1924 railways were carrying 75% of the nation's agricultural produce; in 1938 the figure was 40%. At a local level the wider market opportunities that the branch line gave to farmers was not exploited. The Herriard Estate continued as before to sell its produce at Basingstoke market and not further afield.

To a degree the Southern Railway could have helped to offset some of the losses, but its corporate vision was on wider issues such as unification and the electrification of the system. The company had an efficient dynamic management who pursued a policy of improving those services that would maximise receipts. Byways tended to be ignored, or by the 1930's, closed. It was unfortunate that the line was always run as a self contained branch with no through trains and poor connections at either end. An attempt could have been made to open additional stopping places as unstaffed halts and to utilise railcars but overall, the Southern did not capitalise on the line's light railway status. Even if positive measures had been adopted it is doubtful whether the line would have survived long into nationalisation.

Recognising the threats posed by bus services, the Southern invested in local bus companies. Consequently when the line closed they supported a

A view looking towards Alton showing the position where the line was severed in 1933. Alton Park platform and the power house chimney can be seen behind the tree. The stop block looks as though it has stopped more than enough traffic!

The 'Alton bay', now platform, 1a still in use at Basingstoke some 40 years after the last passenger train left to the destination that gave the platform its name. The new corporate British Rail identity has been applied but the track and buffer stop are from a different age.

Relic re-cycled; a cast iron sign is put to a new use by the Herriard Estate. The sign is of LSWR style but with the new owner, Southern Railway in place of the old pre-grouping company. A memory of the re-instatement of 1924.

request for Aldershot and District Omnibus Company to run a competing service to the Venture operation. This was refused by the Southern Area Traffic Commissioners who saw the willingness of the Southern Railway to promote a bus, but not to provide a train service, as a distinct peculiarity. In many ways the branch was a precursor of what was to follow with buses dominating rural transport. Now the wheel has moved on again and those same rural bus routes have largely gone - the victim of ever changing transport needs.

What of the railway today? Traces of the line can still be seen although much of the trackbed has been incorporated into the surrounding fields. Near Alton the creation of a deer park meant that the undergrowth and trees had to be cleared, thus revealing the shallow embankment at that point. The structures on the line have survived the test of time well. The cottages and Station Masters' houses built by Firbank are all inhabited whilst the corrugated iron station at Bentworth and Lasham still stands - just. The site was used by a coal merchant until recently. At Herriard the concrete faced platforms survive as does a gate post from the level crossing over Bagmore Lane. Cliddesden station has nearly disappeared, the site having been levelled and the platform partially demolished. However in the garden of the stationmaster's house where Mr. Bushnell once lived, the water tower still stands. Alton station is largely unchanged. The platform from whence the branch line train departed still serves a useful purpose; today steam trains over the Mid Hants "Watercress Line" use it to commence their journey to Alresford. The "Alton Bay" at Basingstoke remained in occasional use up to the late 1970's for storing vans or for starting local trains, but it has now gone. Where Viables Lane crossed the line a memento in the form of a section of track has been placed in the middle of the roundabout which now occupies the site. Otherwise much has gone although the observant will still see the occasional bridge abutment or concrete post from a former level crossing.

Perhaps the final words should be reserved for Edward Griffith whose early work was so important. He recounted the tale of a man travelling by rail from Northamptonshire in the 1950's to visit his son at Lasham, who informed his offspring that he would be arriving by train at Bentworth and Lasham station at 10.29 a.m. It was possible 20 years before, but obviously the news of closure had yet to reach Northamptonshire. That is immortality!

APPENDIX

List of Bridges

Number & Name	Location	Construction	Size	Height
Soffitt				
No. 0 Culvert	48m 51ch	Brick	4'	4'
No. 1 Deep Lane	48m 55ch	Brick	12'	14' 3"
No 2 Worting Road	48m 65ch	Brick	15'	14' 5"
No. 3 Winchester Road	49m 26½ch	Brick	30' skew	16' 4"
No. 3A By-Pass bridge				
No. 4 Cliddesden Road	50m 25½ch	Brick	25' skew	17' 6"
No. 5 Swallick	51m 47ch	Brick	12'	13' 8"
No. 6 Winslade	52m 20ch	Brick	12'	14' 3"
No. 7 Shalden	58m 37ch	Brick	12' skew	
No. 8 Dips Road	60m 10ch	Brick	25' skew	16' 6"
No. 9 White Down	60m 22ch	Brick	15' skew	15' 3"
No. 10 Shalden Park	60m 53ch	Brick	20' skew	15' 0"

List of Level Crossings - those prior to 53m 48 ½ch, (mid way between Cliddesden and Herriard are not included in official listings).

No. 1	Jervoise Occupation	53m 48 ½ch
No. 2	Grange Road Public	
No. 3	Bagmore Lane	54m 21 ch
No. 4	Bulls Farm Occupation	54m 55 ½ch
No. 5	Herriard Common	55m 11ch
No. 6	Salter Hatch	55m 71 ¼ch
No. 7	Station Road, (Bentworth)	57m 1 ½ch
No. 8	Brooks Occupation	57 m 70 ½ch
No. 9	Shalden	58m ?ch

Official records of 1901 refer to a total of 41 level crossings:

9 footpaths
20 occupation
11 parish or public road
1 new road

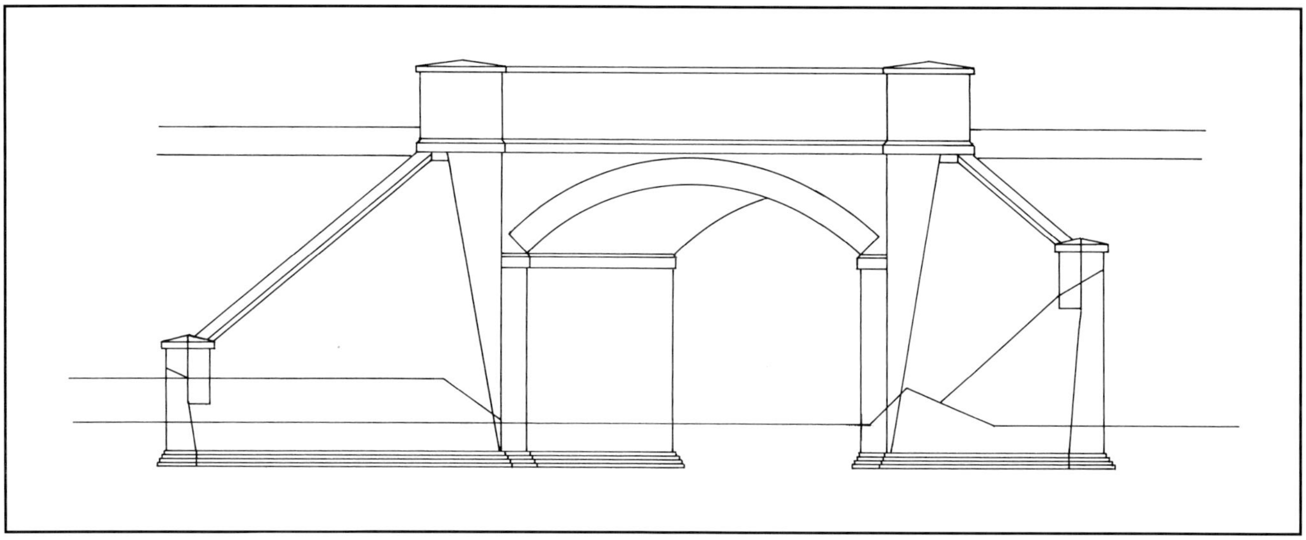

Bridge No.3 – Winchester Road, Basingstoke.

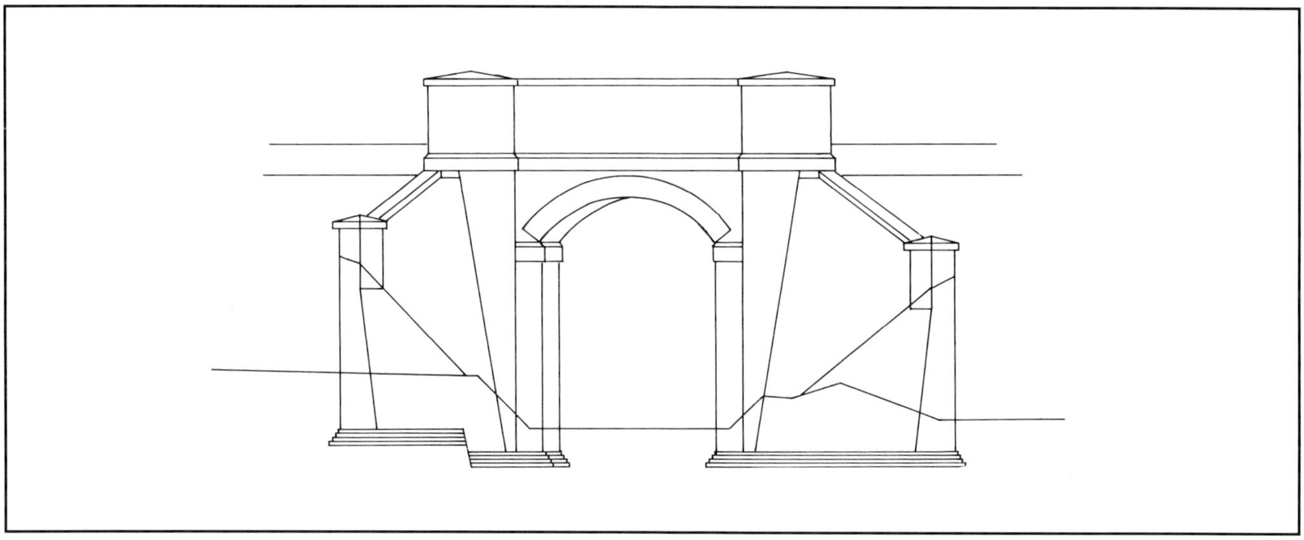

Bridge No.5 – Swallick.

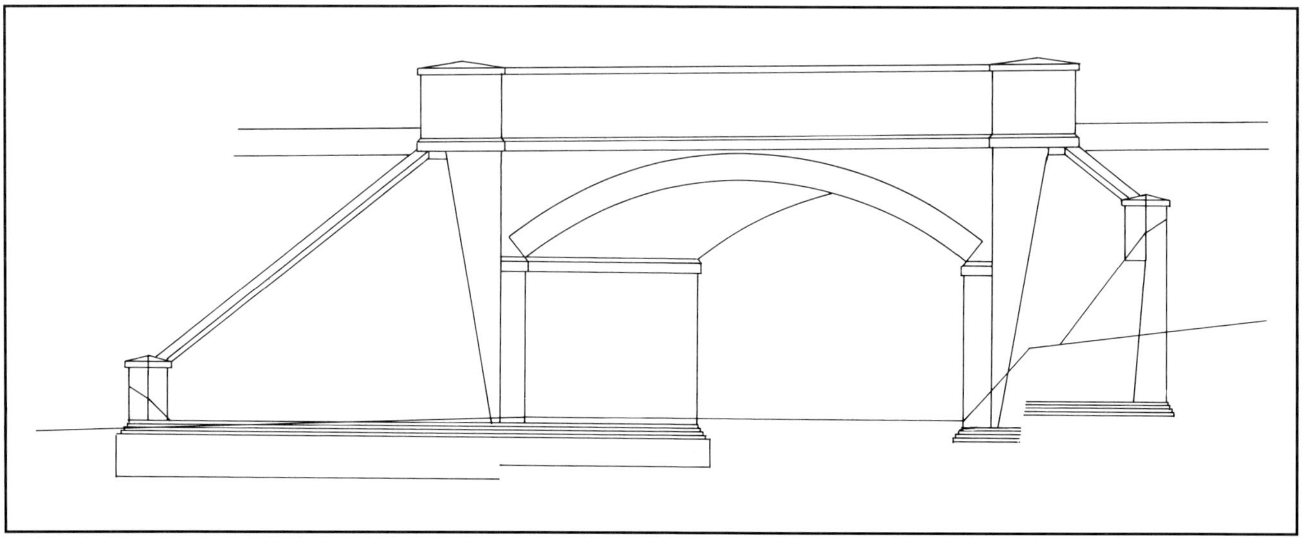

Bridge No.10 – Shalden Park Road, Alton.

ACKNOWLEDGEMENTS

It is always a pleasure in recording the acknowledgements for a book as it is very true to say that without the assistance of the below named none of this could have been written.

Accordingly we would like to thank the following:-

James Adams

David Banks

Eric and Ian Best

British Rail

John Britton

Lt. Col, Ron Burnett

H.C and R.M. Casserley

Denis Cullum

Trefor David

Exeter University Economic History Department

John Fairman

Edward Griffith who despite not being in the best of health freely supported the project and loaned valuable material

Staff at the Hampshire County Museum Service and the Hampshire County Record Office and also various Hampshire Libraries especially Phillipa Stephens from Winchester

Ray Hodson

Richard Hooper

John Huntley of the Huntley Film Archieve

Paul Lacey

Alan Porter

The Keeper and Staff at the Public Record Office

R.C. Riley

Peter Rixon

Ian Shawyer

John Smith (Lens of Sutton)

Dennis Tillman, friends and colleagues of the South Western Circle

Reg, Douglas, Dennis, Derek and Bill - they know who they are and without their help none of this would have been possible.

And especially to Jane and Linda who have been so generous in their support.

BIBLIOGRAPHY

The following books and documents have been consulted:

Hampshire Herald and Alton Gazette
Basingstoke Herald
Hants & Berks Gazette
Documents from Basingstoke UDC & RDC
Documents and Records from the Herriard Estate, courtesy Mr. J. M. Jervoise
Hampshire Chronicle
Hampshire Telegraph
Hansard
Kelly's Directories
Official LSWR/SR record books including:
  Register of LSWR road vehicles
  Minutes of LSWR Directors
  Minutes of LSWR Engineering Committee
  Minutes of LSWR Traffic Committee
  LSWR Staff Records
Minutes and Instructions of Railway Executive Committee
Minutes of Lord Mayor Treloar's Childrens' Hospital and College
The Railway Magazine
The South Western Gazette - various issues
The LSWR staff magazine
The Southern Railway Magazine - various issues
Herepath's Railway Journal
Official Land 'Terrier'
Board of Trade reports
Timetables and Appendices LSWR/SR
Branch Lines to Alton, K. Smith and V. Mitchell
Sir Herbert Walker's Southern Railway, C.F. Klapper
History of the Southern Railway, Dendy Marshall
The Southsea Railway, K. Robertson
LSWR Locomotives, D.L. Bradley
History of the LSWR Vol's. 1 & 2, R.A. Williams
Railway Policy Between the Wars, M.Bonavia
Light Railways, W.J.K. Davies
British Transport, H.J. Dyos and D.H. Aldcroft
Railways and the British Economy, 1830-1914. T.R. Gourvish
Rural England, H. Rider Haggard
Railway Combination and Government 1900-1914, P.J. Cain
The Great Depression in English Agriculture 1872-96, T.W. Fletcher
The Profitability and Performance of British Railways 1870-1914, R.J. Irving.